Make the **Grade** *in GCSE*
Mathematics

TEACH YOURSELF BOOKS

Make the Grade in GCSE Mathematics

Chris and John Shepperd

TEACH YOURSELF BOOKS
Hodder and Stoughton

First published 1987
Second impression 1988

*Copyright © 1987
Chris and John Shepperd*

No part of this publication may be reproduced or transmitted in
any form or by any means, electronically or mechanically,
including photocopying, recording or any information storage
or retrieval system, without either the prior permission in
writing from the publisher or a licence, permitting restricted
copying, issued by the Copyright Licensing Agency,
33–34 Alfred Place, London WC1E 7DP.

British Library Cataloguing in Publication Data

Shepperd, C. J.
  Make the grade in GCSE mathematics.—
  (Teach yourself books)
  1. Mathematics—Examinations, questions, etc.
  I. Title   II. Shepperd, J.A.H.
  510'.76   QA43

ISBN 0 340 40126 5

*Printed in Great Britain for
Hodder and Stoughton Educational,
a division of Hodder and Stoughton Ltd,
Mill Road, Dunton Green, Sevenoaks, Kent,
by Richard Clay Ltd, Bungay, Suffolk*

CONTENTS

Introduction

1. **Numbers** — 1
   1.1 Whole Numbers. 1.2 Decimals. 1.3 Fractions.
   1.4 Sets of Numbers. 1.5 Directed Numbers.
   1.6 Powers and Roots. 1.7 Know your Calculator.
   1.8 Ratio and Proportion. 1.9 Percentages.

2. **Everyday Arithmetic** — 31
   2.1 Buying and Selling. 2.2 Saving and Borrowing.
   2.3 Earning a Wage. 2.4 Reading Tables. 2.5 Time.

3. **Measuring** — 41
   3.1 Units. 3.2 Length. 3.3 Scale. 3.4 Angle.
   3.5 Perimeter. 3.6 Circles. 3.7 Area.
   3.8 Nets and Surface Area. 3.9 Volume.

4. **Geometry** — 71
   4.1 Shape and Symmetry. 4.2 Solids.
   4.3 Constructions and Loci. 4.4 Prove it.

5. **Trigonometry** — 83
   5.1 Trigonometric Ratios. 5.2 Two-dimensional Problems.
   5.3 Three-dimensional Problems. 5.4 Making Waves.

6. **Equations and Graphs** — 95
   6.1 Algebra. 6.2 Factorisation. 6.3 What is a Function?
   6.4 Conversion Graphs. 6.5 Travel Graphs.
   6.6 Plotting Functions. 6.7 Interpreting Graphs.
   6.8 Composite Functions. 6.9 Linear Functions.
   6.10 Linear Equations. 6.11 Problem Solving with Algebra.
   6.12 Simultaneous Equations. 6.13 Quadratic Equations.
   6.14 Linear Inequalities.

7. **Statistics** — 133
   7.1 Collecting and Sorting Data. 7.2 Charts.
   7.3 Pictorial Information. 7.4 Grouped Data.

8. **Probability** — 145
   8.1 Taking a Chance. 8.2 Combined Events.

9 Transformations 151
9.1 Enlarge it. 9.2 Translation, Reflection, Rotation.
9.3 Combined Transformations.

10 Vectors and Matrices 157
10.1 Vectors. 10.2 Matrices. 10.3 Route Matrices.
10.4 Matrix Transformations.

Answers 171

INTRODUCTION

How to use this book

The purpose of this book is to help you to be successful in your mathematics. The material is intended to be used for revision and practice of the work suitable for the General Certificate of Secondary Education (GCSE) examinations in mathematics.

In each section of two or four pages you will find the work based around one main topic. Typically this will include:

1 Brief **notes** on the main ideas, with the key words, in **bold type**, carefully explained.
2 Several **examples** of typical questions followed by worked solutions.
3 Carefully graded **exercises** covering the basic ideas and some longer questions which may require ideas from several sections. All the answers are given at the back of the book.

The intention is that you work through a complete section at one time, jotting down important words, facts, formulae and methods, working through the examples and trying the sample questions. It is not enough to merely read the material—mathematics can only be understood by doing it.

Which paper should I choose?

All systems of GCSE mathematics examinations are in **three levels**, aimed at candidates who are likely to obtain grades as follows:

Level	Intended for candidates likely to obtain grade:	Syllabus content (see p. xi)
Basic, foundation	E F G	List 1
Middle, intermediate	C D E	Lists 1 and 2
Higher, extended	A B C	Lists 1, 2 and further topics

It is important that you enter for the correct level for your ability. You should take the advice of your teachers. You may be tempted to enter for a higher level than advised. If you do, you will probably obtain a worse result because the work will be unsuitable for you.

The contents of the lower and middle level syllabuses are more or less fixed by the National Criteria for mathematics, shown on pages xi–xiii. For the higher level, the syllabus content may vary between the five Examining Boards, and the table on pages xii–xiii shows which topics have been included in your syllabus. Examples and questions marked in this book with an asterisk (*) are usually aimed at the higher level and so may be omitted by those aiming for the middle or lower levels.

This book covers all three levels and all syllabuses, and so you may not need some of the material in the book since it is not present in your syllabus. You must find out the syllabus details of your particular examination and select the correct material from this book.

Make sure that you:
- take note of your teachers' advice and enter for the **correct level** for you;
- are familiar with the **syllabus** details for your level;
- know thoroughly the **basic facts** required so that you feel confident;
- **practise many questions**: the only way to understand and master mathematics is by actually doing it;
- really **get to know your calculator** by regular practice so that you can use it automatically;
- **select the correct parts** of this book to work through;
- **revise** in a sensible way according to a programme: do not leave things to the last minute or you will not get it done and panic.

How do I tackle the examination itself?

Rules
Question papers may differ in their rules (rubric). For example:

Answer **all** questions from Section A and **three** questions from Section B.
Write your answers in the spaces provided.
Write your answers on lined paper unless instructed otherwise.
Answer question 4 on squared paper and attach securely at the back of your answer booklet.

Make sure that you know and understand the rules for your question papers before the day of the examination. On the day, remind yourself of the rules by carefully reading the rubric printed on the examination paper.

Answering questions
Read each question **carefully**, making sure that you understand what is required. When you think you have finished a question, read it again to check that you have answered **all** the question.

Set out your work in an **orderly manner** so that it is clear to you, and to the reader, what you are doing. Marks are given for the correct answer, but also for a correct method and it is important to show this clearly.

When you finish a question, **state the answer clearly**, checking that this is what was wanted. Make sure that the correct units are included and an appropriate level of accuracy stated, when needed.

Introduction

Always **check** your answers – by working backwards from them or by putting the values obtained into the original problem or equation where possible. Look at the answers you have written down and ask yourself whether they are sensible. If you spot a silly answer, check again and correct the error. If you get stuck on a question, move on to another and return to the question later. You may well find that the correct route through to the solution is clearer after concentrating on something different.

Calculators

Many examination papers will expect the confident use of a calculator. **Practise regularly beforehand** with the calculator which you plan to use in the examination, so that you know what functions it can perform. Make sure you have some spare batteries!

Many candidates leave out the explanation when using a calculator. If you only write down a numerical answer after a long calculation and it is wrong, you can get no credit. But if you have **shown all the method** and clearly laid out all the steps made, you may well get most of the marks even if there is a mistake in the answer.

Calculators often give a large number of figures in an answer. Round this answer off to the number of figures asked for, or to a **sensible number of figures** (usually the same number of significant figures as given in the original information). As a check, it is worth doing the calculation roughly with just one-figure accuracy to see if the answer is about right.

Finally, don't exhaust yourself with too much revision just before the day of the examination. Make sure that you get plenty of sleep so you are fresh, alert and ready to give your best on the day of the examination.

What will the coursework involve?

Some of the things to be assessed in GCSE mathematics are not easily tested in a final written examination paper. They will therefore be tested by means of coursework, representing up to thirty per cent of your final marks. Examples of abilities which need to be tested this way are:

oral work and mental work;
practical work;
extended pieces of work;
group work;
investigational work.

Coursework will include a mixture of these activities and will be a different sort of test to the familiar written paper. The work will be done in class with your teacher, who makes the assessment, so it is not something you can easily do by yourself. However, the following notes may help you to understand what your teacher will be looking for.

Introduction x

Oral and mental work
Here you will talk about your mathematics, discussing ideas with your teacher and fellow students, and carry out mental calculations. This may help you to use your mathematical skills in everyday life. Be prepared to talk mathematics: you may well find it easier than writing down the same work, and it will help you to sort out your ideas.

Practical work
This involves using apparatus, weighing, measuring, constructing models and manipulating mathematical shapes. In order to solve a typical problem you might have to read a scale, measure several lengths or angles, make drawings, cut and fold paper or card, look up information or organise results in a table. You can help yourself here by looking for everyday activities which can be dealt with mathematically. Decision-making can often be clarified by a mathematical approach to the problem.

Extended pieces of work
A topic is developed over a period of time, at least one week. It will involve some of the previous activities, discussion, practical work and investigation. Your results may be presented in several ways: a poster, a cassette tape explaining what you discovered, or a model, say.

Group work
In an activity, you may be working on your own or you may need to work in a group with your friends. You must learn how to work together and how to share out the work so that the result is achieved quickly and efficiently.

Investigations
In an investigation, you are not told what to do in detail but are given (or you choose) some idea which can be developed mathematically.
How do you go about this? Here are some tips to remember.

Try some simple cases. The given problem may involve a number of objects and you may be able to find the method first by using a smaller number of objects. Simplify the problem by omitting some of the factors and then see later what effect these factors have when they are reintroduced.

Be systematic. In trying out various possibilities, list them in order of increasing complication and carefully record what you are doing. A table of results will often be useful.

Find a helpful representation. You will need to change the problem into mathematics – this is called choosing a mathematical model. You may need to experiment to find a model which takes all possibilities into account. Try to

xi *Introduction*

use mathematics to simplify your model and then see what this means for your problem.

Look for a pattern. Your examples using small numbers may suggest a pattern which can be extended to more complicated cases. Pattern and structure are most important in mathematics. Often your problem can be solved by recognising a familiar pattern which comes from the solution to another problem. Applying this knowledge to your problem may help towards a solution.

Find a rule and then **check it.** The pattern may help you find a rule which can then be applied generally. Check it fits all the simple cases and then apply your rule to the more difficult cases and see if it still works.

Change the problem to extend the idea. You have solved the original problem. The next thing that a mathematician does is to say: 'What if . . . ?' Try some variations or new conditions and see if you can extend the results.

And finally, **enjoy yourself.** Mathematical investigation is great fun.

Syllabus Comparison: Topics given by Lists 1 and 2

LIST 1 – Basic level
Whole number, factors, multiples
Idea of square root
Directed numbers in practice
4 rules for decimals, percentages
Add and subtract fractions
Equivalence of fractions, decimals and percentages
Percentage of a sum of money
Use of an electronic calculator, checks of accuracy
Estimation, reasonable approximation and rounding
Metric and common units of measurement
Personal and household finance, business transactions
Tables, charts, clocks, dials, scales, maps, plans
Ratio, direct/inverse proportion, measures of rate

Cartesian coordinates and graphs
Measure length/angle. Point, line, angle. Angles at a point
Symmetry and angle properties of triangles and quadrilaterals
Perimeter of rectangle, triangle
Circumference of circle
Area of rectangle, triangle
Simple solid figures, nets
Volume of cuboid
Use of geometrical instruments, scale drawings, bearings
Similarity and enlargement
Symmetry, reflections, rotations
Generalised numbers, substitution
Use of statistics in the media
Data collection, tables, graph, pie-chart, barchart, pictogram
Averages, mean, mode
Probabilities, single events

Introduction xii

LIST 2 – Middle level

Common factors, common multiples
Square roots
Real numbers, standard form
4 rules for fractions, directed numbers
Conversion between fractions, decimals and percentages
Percentage/fraction of one quantity of another, percentage change
Approximate to n significant figures or decimal places
Appropriate limits of accuracy
Interest, appreciation and depreciation
Proportional division
Angles within parallel lines
Symmetry and angle properties of polygons
Circles, tangent-radius property, angle in semicircle
Area of parallelogram, circle
Volume of cylinder
Constructions using simple locus properties
Pythagoras and 3 trigonometrical ratios in 2-dimensional problems
Congruence. Recognition of congruent figures
Algebraic formulae, expansions, factorisation, transforming formulae
Positive and negative integral indices
Function given by a formula, a table of values, a graph
Solution of equations by graphical and numerical methods
Simple linear equations in 1 unknown
Idea of gradient
Construct piecharts and barcharts
Discrete data, frequency distributions, means, mode, median
Equal interval histogram
Simple combined probabilities

Note: The five GCSE Examining Boards are:
LEAG London and East Anglian Group
MEG Midland Examining Group
NEA Northern Examining Association
SEG Southern Examining Group
WJEC Welsh Joint Education Committee
Syllabus topics are also specified by the SMP (Schools Mathematics Project).

The table on page xiii shows the **most common topics not in lists 1 and 2**, as specified by each of the five boards.

Introduction

The number code shows the level at which the Examining Board includes the topic.
1 = Basic, 2 = Middle, 3 = Higher level.

Topic	LEAG	MEG	NEA	SEG	WJEC	SMP
Use of sets and Venn diagrams	3	1	2	3	3	1
Use of set notation and symbols	3		3	3	3	3
Simple number patterns or sequences	1	1	1	1	2	1
Harder angle properties of circles	3		3	3	3	
Length of circular arc, area of circular sector	3	3	2			
Surface area of cylinder, cone, sphere, prism		3	3	3	3	3
Area of trapezium, volume of prism	2	2	3	3	2	3
Volume of pyramid, cone, sphere	3	3	3	3	3	3
Harder geometrical constructions	2		2			2
Pythagoras and trig ratios in 3-dimensions	3	3	3	3	3	3
Length, area and volume of similar figures	3	3	3	3	3	3
Enlargements with negative scale factor	3	3	3	3		3
Data and method for checking congruence	2	2	3			
Translations, simple reflections and rotations	1	2	2	1	2	2
General reflections and rotations		3	3	3	2	2
Combined, inverse transformations	3	3	3			3
Factorise quadratic, difference of 2 squares		3	3	3	3	3
Fractional indices		3	3	3	3	
Notation f(x) for the image of x under f			3	3	3	3
Transformation of harder formulae	3	3		3		
Functions: quadratic, cubic, $a\sqrt{x}$, ax^{-2}, a^x		3	3		3	
Combined, composite and inverse functions			3		3	3
Quadratic functions and quadratic equations	3	3	3	3	3	3
Simultaneous linear equations in 2 unknowns	3	3	3	3	3	
Graphical solution of 2 equations, one linear	3		3			
Gradient as a measure of rate of change	3	2	2	3	3	2
Interpretation of $y = mx + c$	2	3	3	3	3	3
Gradient of tangent, by drawing and calculation	3	2	3	3	3	3
Interpret and estimate area under a graph	3	3	3	2	3	3
Solution set of linear inequality in one variable		3	3	2		3
Linear inequalities in 2 variables			3	3		3
Vectors, addition, product by scalar	3	3	3	3	3	2
Column vectors, unit vectors	3	3	3	3	3	2
Use of vectors to investigate simple shapes		3	3		3	
Matrix representation of data, route matrix	3	3	3			3
Matrix algebra, transformation matrices in 2D	3	3	3	3		3
Grouped data, frequency tables, mode, median	2	3	2	2		3
Cumulative frequency diagram, mean, median	2	3	3			3
Measures of spread, interquartile range		3	3			3
Probabilities associated with AND, OR	3	3	3	3		3

NUMBER
1.1 Whole Numbers

Factors
A **whole** number is a non-zero natural number, so the set of whole numbers is $\{1, 2, 3, \ldots\}$.
A **factor** is a whole number which divides into the given number:
$12 \div 3 = 4$ or $12 = 3 \times 4$ so 3 and 4 are **factors** of 12.
The **set of factors** of 18 is $\{1, 2, 3, 6, 9, 18\}$.
Clearly 1 and the number itself will always be factors of any number.

Prime numbers
A whole number which has only 2 factors, namely 1 and itself, is **prime**.
2, 3, 5, 7, 11 are the first 5 prime numbers (1 is **not** prime by definition).

Rectangle and square numbers
Every whole number which is not prime is a **rectangle** number and this means that it has at least one pair of factors other than 1 and itself.

$28 = 4 \times 7$ so 28 is a rectangle number

$29 = 29 \times 1$ is prime

If a rectangle number can be written as the product of two identical factors, like $36 = 6 \times 6$, then the number is a **square** number as well.

$1 = 1 \times 1, \quad 4 = 2 \times 2, \quad 9 = 3 \times 3$
are the first 3 square numbers.

Factorisation
Every rectangle number can be **factorised**, or **decomposed** into a product of prime factors as follows:

$30 = 2 \times 15$ $455 = 5 \times 91$ $54 = 6 \times 9$
$\quad = 2 \times 3 \times 5$ $\quad = 5 \times 7 \times 13$ $\quad = 2 \times 3 \times 3 \times 3$
                                                                                                                                  $= 2 \times 3^3$

We use power notation to simplify the final factorised form.

Multiples
A **multiple** of a whole number is found by multiplying by any whole number:
$7 \times 3 = 21$ and $7 \times 4 = 28$ so 21 and 28 are **multiples** of 7
A number is, by definition, a multiple of itself, $7 = 7 \times 1$.
The **set of multiples** of 5 less than 30 is $\{5, 10, 15, 20, 25\}$.
Any set of numbers has a **lowest common multiple** (LCM) and a **highest common factor** (HCF).

Example Find the HCF and the LCM of the numbers 12 and 42.

Whole Numbers 2

Factorise 12 and 42 into prime factors. $12 = 2^2 \times 3$, $42 = 2 \times 3 \times 7$.
The HCF is $2 \times 3 = \mathbf{6}$ (using the intersection of the sets of factors).
Write down the first few multiples of 12 and 42.
$\{12, 24, 36, 48, 60, 72, 84, \ldots\}$ $\{42, 84, 126, \ldots\}$
The LCM is the least member of the intersection of these sets **84**.

Example Given the set S, where $S = \{27, 35, 36, 42, 59, 65\}$, write down:
(a) the prime number, (b) the square number, (c) the highest common factor of the two even numbers, (d) the set of factors of the largest number.

(a) **59** is the prime number.
(b) **36** $(= 6 \times 6)$ is the square number.
(c) $36 = 2 \times 18 = 2 \times 2 \times 9 = 2 \times 2 \times 3 \times 3$, $42 = 2 \times 21 = 2 \times 3 \times 7$, so the HCF is $2 \times 3 = \mathbf{6}$.
(d) $65 = 5 \times 13$, so the set of factors of 65 is $\{\mathbf{1, 5, 13, 65}\}$.

Number patterns
Some ordered sets of numbers make interesting patterns:
Odd numbers   1 3 5 7 9 11 13 ...
Even numbers   2 4 6 8 10 12 14 ...
Square numbers   1 4 9 16 25 36 49 ...
Prime numbers   2 3 5 7 11 13 17 ...
Triangle numbers   1 3 6 10 15 21 28 ...
                         triangle patterns of dots
Fibonnacci sequence 1 1 2 3 5 8 13 ... each
                         term is the sum of the preceding two terms.

SAMPLE QUESTIONS
1 Find the set of factors of (a) 24, (b) 45, (c) 56, (d) 75.
2 Expand by multiplication (a) $2^2 \times 3 \times 5$, (b) $7^2 \times 3$, (c) 5^3.
3 Express as a product of prime factors (a) 48, (b) 80, (c) 120.
4 Find the LCM and HCF of (a) 25 and 35, (b) 32 and 48 and 56.
5 Find the square number below 50 which has 5 different pairs of factors.
6 Find the number below 100 which has the largest set of factors.
7 Spot the pattern and write down the next 3 terms in the sequence,
    (a) 2 4 8 16 32 (b) 35 32 29 26 23 (c) 21 26 31 36 41,
    (d) 4 7 11 18 29, (e) 3 6 12 24 48, (f) 1 5 11 19 29,
    (g) 11 121 1331 14 641, (h) 9999 9210 8421 7632.
8 The number sequence 1, 5, 9, 13 ... 29 has 8 terms.
    (a) How many terms are prime? (b) How many terms are square numbers? Find the: (c) LCM of the 3rd and 6th terms, (d) HCF of the 2nd and 7th terms.
*9 $\mathscr{E} = \{$whole numbers between 10 and 20$\}$ (see p. 11). T and F are subsets of $\mathscr{E}$, $T = \{$multiples of 3$\}$, $F = \{$factors of 60$\}$. Draw a Venn diagram showing $\mathscr{E}$, T and F and the position of each member of these sets. Find:
    (a) $T \cap F$, (b) n $(T \cap F')$, (c) the number of primes in the set $(T \cup F)'$.

NUMBER
1.2 Decimals

Decimal (base 10) numbers are written using a **place value** system based on powers of **ten**. For example 2046.03 means two thousands (10^3), four tens (10^1), six units (10^0) and three hundredths (10^{-2}).

10^4	10^3	10^2	10^1	10^0	10^{-1}	10^{-2}	10^{-3}
TTh	Th	H	T	U	t	h	th
	2	0	4	6 .	0	3	

TTh tens of thousands
Th thousands
h hundreds
T tens
U units

t tenths   h hundredths   th thousandths

Addition and subtraction

When decimals are added or subtracted without a calculator the corresponding place values must be aligned, in other words, the digits must be placed in their correct columns.

Example Find (a) $43.72 + 5.6 + 27$ (b) $36 - 19.24$

Write down the sum in columns, fill out with zeros and decimal points where necessary, and then add or subtract as usual.

(a)

	T	U · t	h
	4	3 · 7	2
		5 · 6	
+	2	7 ·	
	7	**6 · 3**	**2**

(b)

	T	U · t	h
	3	6 · 0	0
−	1	9 · 2	4
	1	**6 · 7**	**6**

Multiplying and dividing by powers of 10

Example Find (a) 2046.03×10 (b) 3.14×1000 (c) $1040 \div 100\,000$

(a) Multiplication by 10 will change 2000 into 20 000, 40 into 400, 6 into 60, 0.03 into 0.3. The effect is to move every digit one place to the left into the next highest place value: $2046.03 \times 10 = \mathbf{20\,460.3}$

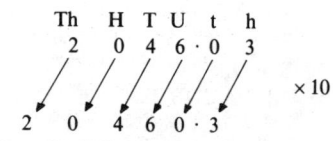

```
Th  H  T  U  t  h
 2  0  4  6· 0  3
                    × 10
 2  0  4  6  0· 3
```

Note: the decimal point remains **fixed**.

(b) $1000 = 10 \times 10 \times 10 = 10^3$ and so multiplication by 1000 will move every digit 3 places to the left. $3.14 \times 1000 = \mathbf{3140}$ (an extra zero is inserted in the units column).

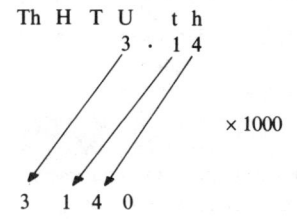

```
Th  H  T  U  t  h
          3· 1  4
                    × 1000
 3  1  4  0
```

Decimals 4

(c) Division by a power of 10 involves moving the digits to the right. $100\,000 = 10^5$ and so every digit moves five places to the right.
$1040 \div 100\,000 = \mathbf{0.0104}$
(extra zeros are used to fill the units and tenths column, trailing zeros may be omitted).

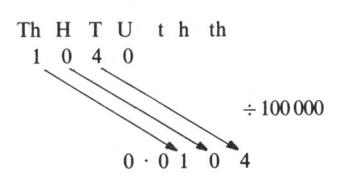

```
Th  H  T  U  t  h  th
    1  0  4  0
                           ÷ 100 000
         0 · 0  1  0  4
```

Note: When large numbers of digits are written it is normal to group them in 3's to make them easier to recognise. 56 million is written 56 000 000.

Multiplication and division

When decimals are to be multiplied or divided without a calculator it is usual to calculate without the decimal points and adjust the answer afterwards.

Example Calculate (a) 6.48×4.3 (b) $7.53 \div 0.03$

(a) 6.48×4.3

$= \mathbf{27.864}$

TTh	Th	H	T	U
		6	4	8
	×		4	3
	1	9	4	4
2	5	9	2	0
2	7	8	6	4

The arrows count the powers of ten, $(\times 100) \times (\times 10) = (\times 1000)$ so the final answer must be divided by 1000.

(b) $7.53 \div 0.03$

$\quad \times 100$

$= 753 \div 3$
$= \mathbf{251}$

Both numbers have been multiplied by 100 and so the answer will be the same.

Rounding

Numbers occurring in mathematical problems may be whole numbers, fractions, decimals, percentages or irrationals. In practice, in order to carry out the calculation, the numbers must often be **rounded**, or changed into decimals of finite length. When an irrational number or a fraction is changed into a decimal, the decimal may need to be rounded either to a fixed number of **decimal places** (**DP**) or to a fixed number of **significant figures** (**SF**).

For instance, $7326.65 = 7300$ (2SF) = 7326.7 (1DP)
$0.0468 = 0.047$ (2SF) = 0.05 (2DP)

Example Express the following as decimals accurate   (a) to 3 significant figures (3SF) and   (b) to 3 decimal places (3DP):
(i) 13/8   (ii) 3/11   (iii) $5\frac{3}{4}$%   (iv) $\sqrt{50}$ (v) pi

5 Decimals

(i) 13/8 means '13 divided by 8' and so may be found by division to be 1.625, a finite decimal. Rounding gives 13/8 = **1.63 (3SF)**, **1.625 (3DP)**.
(ii) Dividing 3 by 11 we find 3/11 = 0.272727 ... where the '27' pattern recurs infinitely. Rounding gives 3/11 = **0.273 (3SF and 3DP)**.
(iii) $5\frac{3}{4}\%$ means '$5\frac{3}{4}$ out of 100' or '5.75 divided by 100', so on division the decimal form is **0.0575** which is accurate to 3 significant figures already. To 3 decimal places $5\frac{3}{4}\% = $ **0.058**.
(iv) The square root of 50 may be found by using tables, a calculator or by trial and error as shown in Section 1.6. Using a calculator, $\sqrt{50} = 7.0710678 = $ **7.07 (3SF)** and **7.071 (3DP)**.
(v) Using tables or a calculator, a decimal approximation for pi is 3.14159 which may be rounded to **3.14 (3SF)** and **3.142 (3DP)**.

Standard form

When very large or very small numbers are written, a **standard form** is used. The mean distance between the Sun and the Earth is 149 500 000 kilometres, accurate to 4 significant figures:

$$149\,500\,000 = 1.495 \times 100\,000\,000 = 1.495 \times 10^8.$$

The charge on an electron is 0.000 000 000 000 000 000 16 coulomb, accurate to 2 significant figures:

$$0.000\,000\,000\,000\,000\,000\,16 = 1.6 \div 10^{19} = 1.6 \times 10^{-19}.$$

The standard form is $a \times 10^n$ where a is a number between 1 and 10 and n is the power of 10, known as the **exponent**.

Calculation in standard form

The rules for manipulating decimals are used together with the rules for combining powers (see page 16) and care must be taken about the placing of the decimal point.

Example Given $p = 3.48 \times 10^4$, $q = 8.02 \times 10^6$, $r = 1.25 \times 10^{-3}$ Calculate
(a) $p+q$, (b) pq, (c) rq, (d) q^2/r, giving the answers in standard form.

(a) To add or subtract standard form numbers align digits of the same place value. $(3.48 \times 10^4) + (8.02 \times 10^6) = 34\,800 + 8\,020\,000 = 8\,054\,800$ which can be written **8.0548×10^6** in standard form.
(b) To multiply or divide standard form numbers separate the mantissa and exponent parts of the numbers and combine them separately.
$(3.48 \times 10^4) \times (8.02 \times 10^6) = (3.48 \times 8.02) \times (10^4 \times 10^6)$
$= 27.9096 \times 10^{10} = \mathbf{2.79096 \times 10^{11}}$
(c) $(1.25 \times 10^{-3}) \times (8.02 \times 10^6) = (1.25 \times 8.02) \times (10^{-3} \times 10^6)$
$10.025 \times 10^3 = \mathbf{1.0025 \times 10^4}$
(d) $(8.02 \times 10^6)^2 \div (1.25 \times 10^{-3}) = (8.02^2 \div 1.25) \times (10^6)^2 \div 10^{-3}$
$= 51.456\,32 \times 10^{15}$
$= \mathbf{5.145\,632 \times 10^{16}}$

Decimals 6

SAMPLE QUESTIONS
1. Write in words the place value of the digit 4 in:
   (a) 24 000, (b) 0.0453, (c) 347 592 000, (d) 0.372 584 321.
2. Use the first result in each case to find the answers to the following questions:
   (a) $23 \times 67 = 1541$, $2.3 \times 6.7 =$; (b) $589 \times 396 = 233\,244$, $5.89 \times 0.0396 =$; (c) $2403 \div 27 = 89$, $2.403 \div 2.7 =$;
   (d) $36\,352 \div 568 = 64$, $363\,520 \div 0.0568 =$.
3. George had seven pairs of mice and each pair produced nine young mice. How many mice has George now?
4. Catherine went shopping with £100 and spent £3.25, £19.99, £12.73, 49p, £10, 78p, £50.03. How much did she spend and how much was left?
5. A football club buys an attacker costing £725 000 and a defender costing £1 250 000 and sells a goalkeeper for £985 000. What is the final cost of these transactions?
6. How many payments of £27.50 are needed to pay off a debt of £522.50?
7. A cinema has 12 rows with 36 seats in each row and a further 7 rows with 22 seats per row. How many seats are there in the cinema? The seats in the first 12 rows cost £3.20 and the other 7 rows cost £2.40 per seat. What will be the toal income from these seats if the cinema is full?
8. A case of 24 cans of soup costs £5.52. What is the cost of 1 can?
9. The attendances at a club over 4 weeks are 23 452, 34 249, 17 849 and 29 450. Calculate the total attendance over the four weeks. The ticket receipts for this period are £304 500. Find the average cost per ticket.
10. Find a 4 significant figure approximation to these square roots using a calculator or tables:
    (a) $\sqrt{3}$ (b) $\sqrt{20}$ (c) $\sqrt{99}$ (d) $\sqrt{0.7}$ (e) $\sqrt{1000}$ (f) $\sqrt{0.002}$
11. Change the following fractions into decimals, see page 9, rounding to 2DP where necessary. (a) 5/9, (b) 3/7, (c) 5/12, (d) 39/100.
12. Express these standard form numbers as decimals:
    (a) 5.639×10^3 (b) 2.208×10^{-2} (c) 2.574×10^7 (d) 2.467×10^{-4}
13. Write the following numbers in standard form:
    (a) 0.003 82 (b) 24 956 300 (c) 20 032.849 64 (d) 0.000 038 984
14. Find the value of n in the following:
    (a) $673 = 6.73 \times 10^n$ (b) $0.000\,563 = 5.63 \times 10^n$
    (c) 3 million $= 3 \times 10^n$ (d) $n = 2.387 \times 10^5$
    (e) $0.009\,186 = n \times 10^{-3}$ (f) $17\,294\,386 = n \times 10^7$
15. Calculate the following without a calculator, giving your answer in standard form. Check on a calculator afterwards.
    (a) $23\,740 \times 0.29$ (b) $23\,740 - 0.297$ (c) $0.0038/65$
    (d) $(1.386 \times 10^{-2}) - (3.56 \times 10^{-3})$ (e) $(4.2 \times 10^5)^3$
16. There are thought to be three quarters of a million grains of sand in a cubic foot. Write this number in standard form.
*17. The speed of sound in air is 3.315×10^2 ms^{-1}, see page 103. Find the time taken for an observer to hear an explosion 3 kilometres away.

NUMBER
1.3 Fractions

Parts of a whole
Fractions are commonly used to describe a quantity which is less than some whole unit. For example; a quarter of an hour, one half pint, an eighth of an inch, two thirds full.
The notation involves two numbers:
the **denominator** (bottom number)—this number describes into how many equal parts the whole has been divided;
the **numerator** (top number)—the number of these parts considered.
$\frac{4}{5}$ means four fifths; the 'whole' has been divided into 5 equal parts and we are concerned with 4 of these parts (Figure 1): $\frac{4}{5} = 4 \times (\frac{1}{5})$.

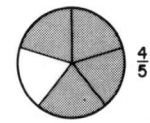

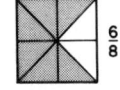

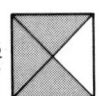

Fig. 1 Fractions.   **Fig. 2** Equivalent fractions

The fraction 'four fifths' is also the result of dividing 4 wholes into 5 equal parts, and fractions are commonly used as a convenient way to describe the result of such a division.

For instance $4 \div 5 = \frac{4}{5}$, sometimes written 4/5.

When the numerator is greater than the denomintor the fraction is called an **improper** fraction and may be rewritten as a **mixed number**, a whole number and a fraction less than 1.

$\frac{8}{3} = \frac{3}{3} + \frac{3}{3} + \frac{2}{3} = 2\frac{2}{3}$ 'Two (whole ones) and two thirds'.

Equivalent fractions
Two fractions may look different but still have the same value and they are then said to be **equivalent**, see Figure 2.

$$\frac{6}{8} = \frac{3 \times 2}{4 \times 2} = \frac{3}{4} \qquad \frac{24}{30} = \frac{2 \times 12}{2 \times 15} = \frac{12}{15} = \frac{3 \times 4}{3 \times 5} = \frac{4}{5}$$

The value of a fraction is unchanged when both its numerator and denominator are multiplied or divided by the same number.

Fractions are usually **cancelled down into simplest form** before being given as answers. This involves finding the highest common factor (HCF) of the numerator and denominator and dividing both by this number.

Example Express in simplest form (a) $\frac{64}{72}$ (b) $169 \div 26$

(a) The highest common factor of 72 and 64 is 8, so $\frac{64}{72} \overset{\div 8}{\underset{\div 8}{=}} \frac{8}{9}$

(b) $169 \div 26 = \frac{169}{26}$, which may be cancelled down to $\frac{13}{2}$ on dividing both top and bottom by 13. $\frac{13}{2}$ may then be written as a mixed number **$6\frac{1}{2}$**

Comparing fractions

To compare fractions, first rewrite them as equivalent fractions with the *same* denominator and then compare the numerators.

Example Place the fractions $\frac{5}{8}$, $\frac{3}{5}$, $\frac{2}{3}$, in order, smallest first.

The lowest common multiple of the denominators 8, 5 and 3 is 120.

$$\frac{5}{8} = \frac{5 \times 15}{8 \times 15} = \frac{75}{120} \qquad \frac{3}{5} = \frac{3 \times 24}{5 \times 24} = \frac{72}{120} \qquad \frac{2}{3} = \frac{2 \times 40}{3 \times 40} = \frac{80}{120}$$

So $\quad \frac{3}{5} < \frac{5}{8} < \frac{2}{3}$

Fractions of a quantity

Example Find (a) $\frac{3}{4}$ of £24, (b) $\frac{7}{10}$ of 3.78 m; (c) What fraction is 12 kg of 96 kg?

(a) $\frac{1}{4}$ of £24 is £24 ÷ 4 which is £6, so $\frac{3}{4}$ of £24 is £6 × 3 which is **£18**.
Alternatively the work may be set out as follows:

$\frac{3}{4}$ of $24 = \frac{3}{4} \times 24 = \dfrac{3 \times 24}{4} = \dfrac{3 \times 6}{1} = 18$, so the answer is **£18**.

(b) $\frac{1}{10}$ of 3.78 m = (3.78 ÷ 10) m = 0.378 m,
so $\frac{7}{10}$ of 3.78 m = (0.378 × 7) m = **2.646 m**

(c) 12 kg is $\frac{12}{96}$ of 96 kg. $\frac{12}{96} = \frac{1}{8}$

Adding and subtracting fractions

Example Find (a) $\frac{3}{7} + \frac{4}{7} + \frac{6}{7}$ (b) $\frac{1}{3} + \frac{2}{5}$ (c) $\frac{9}{4} - \frac{1}{6}$

(a) Fractions with a **common denominator** are added or subtracted directly by **adding or subtracting their numerators**.

$$\frac{3}{7} + \frac{4}{7} + \frac{6}{7} = \frac{3+4+6}{7} = \frac{13}{7} = 1\frac{6}{7}$$

(b) Fractions with **different denominators** must first be **changed into equivalent fractions** with the **same denominator** and then proceed as in (a). The LCM of 3 and 5 is 15 so we work in $\frac{1}{15}$'s, see Figure 3.

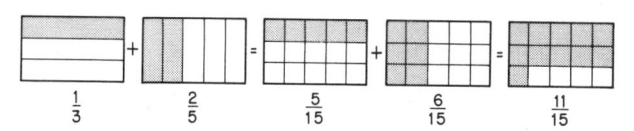

$\frac{1}{3} \qquad \frac{2}{5} \qquad \frac{5}{15} \qquad \frac{6}{15} \qquad \frac{11}{15}$

Fig. 3 Adding fractions with different denominators

(c) The LCM of 4 and 6 is 12 so we work in $\frac{1}{12}$'s here.

$$\frac{9}{4} - \frac{1}{6} = \frac{9 \times 3}{4 \times 3} - \frac{1 \times 2}{6 \times 2} = \frac{27}{12} - \frac{2}{12} = \frac{25}{12} = \mathbf{2\frac{1}{12}}$$

9 Fractions

Multiplying and dividing fractions
Fractions are multiplied by multiplying their corresponding numerators and denominators.
Dividing by a fraction is the same as multiplying by its **reciprocal**.
The reciprocal of $\frac{3}{4}$ is $\frac{4}{3}$, with numerator and denominator reversed.

Example Find (a) $\frac{2}{3} \times \frac{5}{8}$ (b) $\frac{10}{24} \div \frac{5}{6}$

(a) It is advisable to cancel before multiplying where possible.
$$\frac{2}{3} \times \frac{5}{8} = \frac{2 \times 5}{3 \times 8} = \frac{\cancel{2} \times 5}{3 \times 4 \times \cancel{2}} = \frac{1 \times 5}{3 \times 4} = \frac{5}{12}$$

(b) $$\frac{10}{24} \div \frac{5}{6} = \frac{10}{24} \times \frac{6}{5} = \frac{10 \times 6}{24 \times 5} = \frac{2 \times \cancel{5} \times \cancel{6}}{4 \times \cancel{6} \times \cancel{5}} = \frac{2 \times 1}{4 \times 1} = \frac{1 \times 1}{2 \times 1} = \frac{1}{2}$$

Fractions into decimals, and back again
Decimals may be converted into fractions quite easily by using the least place value of the decimal.
$$5.36 = 536 \div 100 = \frac{536}{100} = \frac{134}{25} = 5\tfrac{9}{25}. \quad 0.035 = \frac{35}{1000} = \frac{7}{200}.$$

Example Change into a decimal accurate to 3SF: (a) $\frac{5}{8}$ (b) $\frac{31}{73}$

(a) $\frac{5}{8} = 5 \div 8 = \mathbf{0.625}$ after the division opposite. 8)5.000
  ——————
  0.625

This decimal is already accurate to 3 SF.

(b) $\frac{31}{73} = 31 \div 73 = 0.4246575$ (using a calculator), an infinite recurring decimal, which can be written as **0.425 (3SF)**.

*****Example** A population of Rainbow Green butterflies is estimated to be decreasing by one quarter each year. In 1987 the population numbers 800. How many butterflies are expected to be in the colony by (a) 1988, (b) 1992? What fraction of the original colony of 1987 will remain by 1992?

If the population is decreasing by one quarter then three quarters of the population remain.
So in 1988 the colony will be $\frac{3}{4}$ of $800 = 800 \times \frac{3}{4} = \mathbf{600}$ **butterflies**.
In 1989 there will be $\frac{3}{4}$ of $600 = (800 \times \frac{3}{4}) \times \frac{3}{4} = 800 \times (\frac{3}{4})^2$
In 1992 there will be $800 \times \frac{3}{4} \times \frac{3}{4} \times \frac{3}{4} \times \frac{3}{4} \times \frac{3}{4} = 800 \times (\frac{3}{4})^5 = 189.84$ and the estimate of size of the colony will be **190 butterflies**.
The fraction of the original colony remaining in 1992 is $\frac{190}{800} = \frac{19}{80}$

Fractions 10

SAMPLE QUESTIONS

1. Express the following as fractions in simplest form:
   (a) $\frac{36}{45}$ (b) $\frac{81}{63}$ (c) $\frac{128}{512}$ (d) $\frac{91}{147}$
2. Place the following set of fractions in order, smallest first:
   (a) $\{\frac{4}{7}, \frac{5}{9}\}$ (b) $\{\frac{6}{5}, \frac{11}{9}, \frac{13}{10}\}$ (c) $\{\frac{1}{3}, \frac{4}{9}, \frac{7}{20}, \frac{12}{29}\}$
3. Find (a) $\frac{2}{3}$ of £57 (b) $\frac{5}{9}$ of 3.6 kg (c) $\frac{7}{8}$ of 432 sheep
4. Express the first number as a fraction of the second
   (a) 24, 60 (b) 15, 105 (c) 28, 161 (d) 32, 656
5. Calculate (a) $\frac{3}{4}+\frac{6}{9}$ (b) $\frac{2}{5}-\frac{1}{4}$ (c) $\frac{3}{8}+\frac{1}{4}-\frac{5}{16}$ (d) $3\frac{1}{2}+2\frac{1}{4}$
   (e) $\frac{2}{5} \times \frac{7}{10}$ (f) $\frac{3}{5} \div \frac{2}{7}$ (g) $\frac{12}{25} \times \frac{5}{16}$ (h) $(\frac{2}{3}+\frac{5}{6}) \times \frac{1}{4}$
   (i) $(\frac{3}{7}-\frac{1}{3})/(\frac{2}{3}+\frac{5}{7})$
6. Express as a decimal, accurate to 3 DP:
   (a) $\frac{5}{6}$, (b) $\frac{7}{9}$, (c) $\frac{12}{25}$, (d) $\frac{6}{7}$, (e) $3\frac{7}{8}$, (f) $22 \div 7$.
7. Mary receives £2.40 pocket money a week. She spends one third on magazines and three quarters of the remainder on sweets. How much does Mary spend on magazines, on sweets and how much does she have left?
8. The Ratagonian Airforce began the latest war with 35 planes and 56 helicopters. Losses by the end of the first week were three fifths of the planes and five eighths of the helicopters. How many of each type of aircraft were left after the first week and what fraction of the original total were left after the first week?
*9. A wealthy man distributes £1000 among his four children at Christmas. George receives one quarter, Henry and Jean one fifth each and the favourite, Penelope, receives the rest. How much does each child receive and what fraction does Penelope receive?
*10. A caterpillar, weighing 28.5 g, eats sufficient leaves in one day to increase its weight by one eighth. Find its new weight. If it eats enough the following day to increase its new weight by one tenth find its weight after the second day. By what fraction has the caterpillar increased its weight over the 2 days?
*11. A car, priced at £5600 new, loses one quarter of its value in the first year and then a further fifth of its current value for each of the next 3 years. What is the value of the car after: (a) 1 year, (b) 2 years, (c) 4 years. What fraction of its original value is the car worth after four years?

NUMBER
1.4 Sets of Numbers

The language of sets
Set language and notation is often used in mathematics to describe collections of numbers or points.

***Example** Let $\mathscr{E}$ be the whole numbers less than 10, D be the set of odd numbers less than 10 and S be the set $\{2, 3, 4, 5\}$. Draw a Venn diagram showing the elements of these sets and find (a) $n(S)$, (b) $S \cap D$, (c) $S \cup D$, (d) D', (e) $\{8\} \cap D$

$\mathscr{E}$ is the **universal set**, the largest set currently being considered. In this case $\mathscr{E} = \{1, 2, 3, 4, 5, 6, 7, 8, 9\}$.

Set D is given by a description and the **members** (or **elements**) may be listed, $D = \{1, 3, 5, 7, 9\}$.

The Venn diagram drawn in Figure 1 shows the members of the sets $\mathscr{E}$, D and S placed in their correct positions. So 1 is an element of D, written $1 \in D$, and the number 1 is drawn within the set loop labelled D.

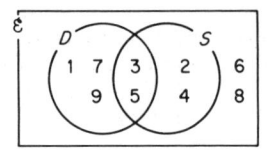

Fig. 1 Venn Diagram

(a) $n(S)$ means 'the number of elements' of S, and so $n(S) = \mathbf{4}$.
(b) $S \cap D$ is the **intersection** of S and D, the set containing elements which are in **S and D**, (the overlapping region), so $S \cap D = \{\mathbf{3, 5}\}$.
(c) $S \cup D$ is the **union** of S and D, the set containing elements which are in **S or D**, so $S \cup D = \{\mathbf{1, 2, 3, 4, 5, 7, 9}\}$.
(d) D' is the **complement** of D, the set of elements in $\mathscr{E}$ which are **not** in D, or outside D, so $D' = \{\mathbf{2, 4, 6, 8}\}$.
(e) $\{8\} \cap D$ is the intersection of the set D and the set containing just one number 8. Since 8 is not an element of D, written $8 \notin D$, there are no elements in this intersection, so $\{8\} \cap D = \boldsymbol{\phi}$, **the empty set**.

Real numbers
The set of numbers normally used in mathematics is called the set of **real numbers**. These numbers can be represented as the points on the **real number line**, $\mathbb{R}$, which has an **origin** O representing the number zero. The value of each number is given by its distance from the origin—**positive** to the right and **negative** to the left. The real numbers are **ordered** naturally, the greatest on the right and the least on the left. For instance, negative one is greater than negative two, written $-1 > -2$.

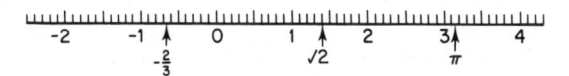

Fig. 2 The real number line

Sets of Numbers 12

The set of real numbers has the following important subsets.

$\mathbb{N}$ the set of **natural** numbers $\quad \{0, 1, 2, 3, 4, \ldots\}$
$\mathbb{Z}$ the set of **integers** $\quad\quad\quad\quad \{\ldots, -3, -2, -1, 0, 1, 2, 3, \ldots\}$
$\mathbb{Q}$ the set of **rational** numbers $\quad \{p/q : p \in \mathbb{Z}, q \in \mathbb{N}, q \neq 0\}$

$\mathbb{Q}$, the set of fractions, contains both $\mathbb{N}$ and $\mathbb{Z}$ as subsets, since a natural number $2 (= 8/4)$ and an integer $-1 (= -3/3)$ are also members of $\mathbb{Q}$. The definition for $\mathbb{Q}$ reads '$\mathbb{Q}$ is the set of numbers p divided by q, where p is an integer and q is a non-zero natural number'.

When all the fractions are inserted on the real number line it might be thought that the line is full with no gaps remaining. But many numbers, such as $\sqrt{2}$ and π, cannot be expressed exactly as fractions. These are all the non-recurring decimal numbers, called the **irrational** numbers.

Example Find the smallest of the sets φ (the empty set), $\mathbb{N}, \mathbb{Z}, \mathbb{Q}, \mathbb{R}$ in which the following numbers, or sets of numbers, lie.
(a) $22/7$ (b) $(-3)^2$ (c) $-12/4$ (d) $\sqrt{3}$ (e) $\sqrt{16}$ (f) $\{x : x^2 = 4\}$
(g) $\{x : x^2 = -4\}$ (h) $\{x : x^2 = 5\}$ (i) 15%

(a) 7 does not divide into 22 exactly so $22/7$ is a rational in $\mathbb{Q}$.
(b) $(-3)^2 = 9$ which is a natural number in $\mathbb{N}$.
(c) $-12/4 = -3$ which is an integer in $\mathbb{Z}$.
(d) $\sqrt{3}$ is a **surd** (square root) which cannot be simplified so is an irrational in $\mathbb{R}$.
(e) $\sqrt{16} = 4$ which is a natural number in $\mathbb{N}$.
(f) The solutions of the equation $x^2 = 4$ are $x = 2$ or -2, so the set containing these two numbers is $\mathbb{Z}$, the set of integers.
(g) This equation has no real solutions so the set is the empty set ϕ.
(h) The solutions of the equation $x^2 = 5$ are $x = \sqrt{5}$ or $-\sqrt{5}$, so the set containing these two numbers is $\mathbb{R}$.
(i) $15\% = 0.15 = 15/100$ which is a rational number in $\mathbb{Q}$.

SAMPLE QUESTIONS
1. Which of the following numbers are irrational: $\sqrt{6400}$, $\sqrt{2.5}$, $2/3$, $2 + \sqrt{3}$?
2. Draw a large diagram showing the part of the real number line between 3.1 and 3.2 and marking the position of the numbers:
   3.1, 3.2, 3.15, 393/125, 22/7, $\sqrt{10}$, π, $1 + 3/\sqrt{2}$.
3. Let S be the set of numbers $\{2, -3, 0.8, \sqrt{2}\}$. Write down
   (a) the least member of S, (b) the number of rational members of S, (c) the member of S which is closest to 'one and five twelfths'.
4. Find whether the statement is true or false:
   (a) $-4 \in \mathbb{Z}$, (b) $\tfrac{12}{4} \in \mathbb{N}$, (c) $\sqrt{8} \in \mathbb{Q}$, (d) $\mathbb{R} \subset \mathbb{Z}$, (e) $3.14 \in \mathbb{Q}$.

NUMBER
1.5 Directed Numbers

The real numbers are made up of **positive** numbers, **negative** numbers and the number zero as shown in Figure 2, page 11. Every non zero real number is a **directed (signed)** number which is either positive or negative, although it is normal to omit the positive sign. It is incorrect to call them 'plus' or 'minus' numbers even though a plus or minus sign is used.

$+5$ is the number 'positive 5' which is the same as the number 5.
-2 is the number 'negative 2' which is the number 2 less than zero.

Fractions, decimals and percentages may all be directed, positive or negative.
The solution to the equation, $3x + 5 = 0$, is $x = -5/3$.
A temperature may be written -2.5 degrees Celsius, 2.5 degrees below zero.
A percentage change may be written -10%, meaning a decrease of 10%.

Ordering
The directed numbers are an **ordered** set and this ordering may be shown by using the symbols $>$ meaning **is greater than** and $<$ meaning **is less than**.
$-2 < +5$ since -2 is to the left of $+5$ on the real number line.

$+7 > +3 \quad -7 < -3 \quad -1/2 < 0.5$ are all true inequalities.

These symbols are often used to describe a subset of the real numbers.

$\{x: -4 < x < 3\}$ is the set of all real numbers between -4 and $+3$, excluding the end-points. If -4 and 3 are also to be included then the $\leq$ symbol, meaning **less than or equal** is used, $\{x: -4 \leq x \leq 3\}$.

Operations
Directed numbers can be combined by the usual arithmetic operations by following the rules below.

Addition and Subtraction
Adding a **Negative** is the same as **Taking** a **Positive**.
Taking a **Negative** is the same as **Adding** a **Positive**.

$3 + -2 = 3 - +2 = 3 - 2 = 1 \qquad -3 + -2 = -3 - +2 = -3 - 2 = -5$
$3 - -2 = 3 + +2 = 3 + 2 = 5 \qquad -3 - -2 = -3 + +2 = -3 + 2 = -1$

Multiplication and Division
Positive × **Positive** = **Positive**; **Negative** × **Negative** = **Positive**,
Positive × **Negative** = **Negative**; **Negative** × **Positive** = **Negative**.
The square of a number is always positive.

$+3 \times +2 = +(3 \times 2) = +6 \qquad -3 \times -2 = +(3 \times 2) = +6$
$+3 \times -2 = -(3 \times 2) = -6 \qquad -3 \times +2 = -(3 \times 2) = -6$

The same rules hold for division.
$+8 \div -4 = -(8 \div 4) = -2 \qquad -8 \div -4 = +(8 \div 4) = +2$

Directed Numbers

Example Calculate (a) $(-3 + -6)^2$ (b) $(6 - (-2 - 4))/(-3)$
*(c) $f(-3)$ where $f(x) = 2x^2 - 5x - 1$

(a) Begin with the expression inside the brackets
$-3 + -6 = -3 - +6 = -3 - 6 = -9$
Squaring a number is the same as multiplying it by itself and so
$(-9)^2 = -9 \times -9 = +(9 \times 9) = \mathbf{81}$

(b) Starting with the top of the fraction
$6 - (-2 - 4) = 6 - (-6) = 6 - -6 = 6 + +6 = 12$
Then the expression can be written $12/(-3) = -(12/3) = \mathbf{-4}$

(c) $f(-3) = 2(-3)^2 - 5(-3) - 1$ (see p. 107)
$= 2(-3 \times -3) - -15 \quad -1$
$= \quad 18 \quad\quad + 15 \quad -1$
$= \mathbf{32}$

SAMPLE QUESTIONS

1. At 6 pm one evening the temperature was 2°C and during the next 6 hours it falls by 8°C. What was the temperature at midnight that night?
2. A Bank shows that a customer's account is overdrawn by printing the balance as a negative number. A woman has £23.56 in her account but then writes a cheque for £46.99. What will her balance show if the cheque is cleared? What will her balance show if she then pays £25 into her account?
3. Find: (a) $-2 - 6 + -4$ (b) $-5 \times -4 \div -2$
   (c) $(-4 + -3)^3$ (d) $5 - (-3 + 7)/(-5)$
   (e) $(8 - -3)^2 - (-8 + 3)^2$
4. Show the given values of x and y are solutions of the equations
   (a) $x = -3$, $3(4 - x) = 24 + x$ (b) $x = 4$ or -1, $x^2 - 3x = 4$
   (c) $x = -2$, $y = 3$, $x - y = -5$, $2x - 3y = -13$ (see p. 122)
*5. Find the solution of the equation $n(n + 2) = 35$ where
   (a) n is a positive integer, (b) n is a negative integer. (see p. 125)
6. $v = u + at$. Find v when (a) $u = 2$, $a = -9$, $t = 3$ (b) $u = -4$, $a = -2$, $t = 5$
*7. $\mathbf{a} = \begin{pmatrix} 2 \\ -2 \\ -5 \end{pmatrix}$ $\mathbf{b} = \begin{pmatrix} 0 \\ 3 \\ -4 \end{pmatrix}$ $\mathbf{c} = \begin{pmatrix} -1 \\ -2 \\ 3 \end{pmatrix}$ Find (a) $\mathbf{a} + \mathbf{b} + \mathbf{c}$,
   (b) $3\mathbf{b} - 2\mathbf{c}$ (see pp. 159–60)
*8. Evaluate the function $f(x)$ for $x = 2, 1, 0, -1, -2$ and -3 where
   (a) $f(x) = x^2 - 3x$ (b) $f(x) = 5 - 4x^2$ (c) $f(x) = 2x(3 - x^2)$
*9. Find the image of $A(0, 2)$, $B(-1, -1)$ and $C(3, -4)$ after transformation by the matrix $\mathbf{M}$ where $\mathbf{M} = \begin{pmatrix} -1 & 2 \\ 1 & -3 \end{pmatrix}$ (see p. 167).

NUMBER
1.6 Powers and Roots

Positive powers

The power notation is used whenever a number or expression is repeatedly multiplied by itself.

$4^2 = 4 \times 4 = 16$ is read '4 **squared**' or '4 **to the power** 2'.
$2^3 = 2 \times 2 \times 2 = 8$ is read '2 **cubed**' or '2 **to the power** 3'.
$7^5 = 7 \times 7 \times 7 \times 7 \times 7 = 16807$ is '**the fifth power of** 7' or '**7 to the power** 5'.

The most common error is to think that 2^3 means $2 \times 3 (= 6)$, rather than 8.

Example Find (a) $5 \times 2^5 + 2 \times 5^2$ (b) the larger of 3^5 and 5^3

(a) Calculation of powers must be made before multiplication, therefore
$5 \times 2^5 + 2 \times 5^2 = 5 \times 32 + 2 \times 25 = 160 + 50 = \mathbf{210}$
(b) $3^5 = 3 \times 3 \times 3 \times 3 \times 3 = 243$, $5^3 = 5 \times 5 \times 5 = 125$ so $\mathbf{3^5}$ is the larger.

Square and cube roots

The inverse operation to **squaring** is **finding the square root of**.
$3^2 = 9$ and so the **square root** of 9 is 3, written $\sqrt{9} = 3$.

Note: $(-3)^2 = 9$ too, but the square root is defined to be positive 3.

Similarly, **finding the cube root of** is the inverse operation to **cubing**.
$5^3 = 125$ and so the **cube root** of 125 is 5, written $\sqrt[3]{125} = 5$.

Example Find (a) $\sqrt{16}$ (b) $\sqrt[3]{27}$ (c) $\sqrt{25} - \sqrt{4} + \sqrt{49}$

(a) $16 = 4 \times 4$, so $\sqrt{16} = \mathbf{4}$ (b) $27 = 3 \times 3 \times 3$, so $\sqrt[3]{27} = \mathbf{3}$
(c) $25 = 5 \times 5$, $4 = 2 \times 2$, and $49 = 7 \times 7$, so
$\sqrt{25} - \sqrt{4} + \sqrt{49} = 5 - 2 + 7 = \mathbf{10}$

Although $\sqrt{9} = 3$, a natural number, in general the result of finding the square root or cube root will be an irrational number.

$\sqrt{2} = 1.4142136\ldots$, $\sqrt{3} = 1.7320508\ldots$, $\sqrt[3]{100} = 4.6415888\ldots$, where the dots mean that the digits carry on with no particular pattern.

We may find decimal approximations to these roots by using a calculator, a table of values or by a decimal search method. On a simple calculator there may be $\boxed{\sqrt{}}$ and $\boxed{\sqrt[3]{}}$ keys.

On a scientific calculator you may need to use the inverse function and the $\boxed{x^2}$ or $\boxed{x^y}$ keys.

Example Write down the calculator key sequence to find: (a) $\sqrt{20}$ (b) $\sqrt[3]{9}$
(c) $(\sqrt{3})^5$ (d) $\sqrt{(3^5)}$ and round to 3 decimal places.

(a) Key in $\boxed{20}$ $\boxed{\sqrt{}}$, and the result should be $4.472136 = \mathbf{4.472}$ **(3DP)**.

Powers and Roots 16

(b) Key in $\boxed{9}$ $\boxed{x^{1/y}}$ $\boxed{3}$ $\boxed{=}$, and the cube root of 9 is given as 2.0800838 = **2.080 (3DP)**,
(on some calculators $\boxed{x^{1/y}} = \boxed{\text{INV}}\;\boxed{x^y}$)

(c) Key in $\boxed{3}$ $\boxed{\sqrt{}}$ $\boxed{x^y}$ $\boxed{5}$ $\boxed{=}$,
giving 15.588457 = **15.588 (3DP)**.

(d) Key in $\boxed{3}$ $\boxed{x^y}$ $\boxed{5}$ $\boxed{=}$ $\boxed{\sqrt{}}$,
also giving 15.588457 = **15.588 (3DP)**.

Decimal search

Most calculators have a square root key but not all have a cube root key. The **decimal search** method may be used to find a cube root and it relies on a type of trial and error approach which gradually focusses in on the correct decimal approximation.

Example Find $\sqrt[3]{28}$, accurate to 3DP, by a decimal search method.

$3^3 = 27$, $3.1^3 = 29.79$, so $\sqrt[3]{28}$ is between 3.0 and 3.1.
$3.05^3 = 28.3726$, $3.03^3 = 27.8181$, $3.04^3 = 28.0945$,
so $\sqrt[3]{28}$ is between 3.03 and 3.04.
$3.035^3 = 27.956$, $3.037^3 = 28.0114$, $3.036^3 = 27.9837$,
so $\sqrt[3]{28}$ is between 3.036 and 3.037.
$3.0365^3 = 27.9975 < 28$, and so $\sqrt[3]{28} = $ **3.037 (3DP)**.

Rules of indices (powers)

The following rules are used to combine powers in order to simplify expressions involving powers of numbers or letters.

1 $a^x \times a^y = a^{x+y}$ **2** $a^x \div a^y = a^{x-y}$ **3** $(a^x)^y = a^{xy}$
4 $a^x \times b^x = (ab)^x$ **5** $a^x \div b^x = (a/b)^x$

These rules may be demonstrated for positive powers as follows.

1 $5^2 \times 5^4 = (5 \times 5) \times (5 \times 5 \times 5 \times 5) = 5 \times 5 \times 5 \times 5 \times 5 \times 5$
$= 5^6 = 5^{2+4}$.

2 $6^5 \div 6^2 = (6 \times 6 \times 6 \times 6 \times 6) \div (6 \times 6) = \dfrac{6 \times 6 \times 6 \times 6 \times 6}{6 \times 6}$

$= 6 \times 6 \times 6 = 6^3 = 6^{5-2}$.

3 $(3^2)^4 = (3 \times 3) \times (3 \times 3) \times (3 \times 3) \times (3 \times 3) = 3^8 = 3^{(2 \times 4)}$.

4 $2^3 \times 5^3 = (2 \times 2 \times 2) \times (5 \times 5 \times 5) = (2 \times 5) \times (2 \times 5) \times (2 \times 5)$
$= (2 \times 5)^3 = 10^3$.

5 $6^4 \div 3^4 = (6 \times 6 \times 6 \times 6) \div (3 \times 3 \times 3 \times 3)$
$= (6/3) \times (6/3) \times (6/3) \times (6/3) = (6/3)^4 = 2^4$.

17 Powers and Roots

Example Simplify the following expressions

(a) $\dfrac{8^5 \times 8^2}{8^3}$ (b) $\dfrac{x^3 x^2}{x^4}$ (c) $\dfrac{(2x^2)^4}{(x^3)^2}$ (d) $(2^5 \times 6^5) \div (4^3 \times 2^4)$

(a) $\dfrac{8^5 \times 8^2}{8^3} = \dfrac{8^{5+2}}{8^3} = \dfrac{8^7}{8^3} = 8^{7-3} = \mathbf{8^4}$

(b) $\dfrac{x^3 x^2}{x^4} = \dfrac{x^{3+2}}{x^4} = \dfrac{x^5}{x^4} = x^{5-4} = x^1 = \mathbf{x}$

(c) $\dfrac{(2x^2)^4}{(x^3)^2} = \dfrac{2^4 (x^2)^4}{x^{3 \times 2}} = \dfrac{16 x^{2 \times 4}}{x^6} = 16 x^{8-6} = \mathbf{16 x^2}$

(d) $(2^5 \times 6^5) \div (4^3 \times 2^4) = (2 \times 6)^5 \div (4^3 \times 4^2)$
    (since $2^4 = 2 \times 2 \times 2 \times 2 = 4 \times 4 = 4^2$),
    $= 12^5 \div 4^5 = (12/4)^5 = \mathbf{3^5}$

Negative powers

The rules of indices for positive powers may be used to extend the meaning of the power notation to negative powers. The operation 'multiply by 4' moves one term up the sequence of powers of 4, and 'divide by 4' moves the opposite way, see Figure 1 below.

4^{-3}	4^{-2}	4^{-1}	4^0	4^1	4^2	4^3	4^4
$\tfrac{1}{64}$	$\tfrac{1}{16}$	$\tfrac{1}{4}$	1	4	16	64	256

(with $\times 4$ arrows between successive terms)

So $4^0 = 1$, and indeed any non zero number to the power 0 is 1, $a^0 = 1$.

$4^{-2} = \dfrac{1}{4^2} = \dfrac{1}{16}$. A number raised to a negative power is equal to the reciprocal of the corresponding positive power. $a^{-x} = \dfrac{1}{a^x}$

Example Find (a) 3^{-2} (b) 7^0, (c) 8^{-3}.

(a) $3^{-2} = \dfrac{1}{3^2} = \dfrac{1}{\mathbf{9}}$. (b) $7^0 = \mathbf{1}$. (c) $8^{-3} = \dfrac{1}{8^3} = \dfrac{1}{\mathbf{512}}$.

Powers and Roots 18

*Fractional powers
Square roots, cube roots and indeed any other roots may be expressed in power notation by extending the rules of indices to fractional powers.

$9^{\frac{1}{2}} \times 9^{\frac{1}{2}} = 9^{(\frac{1}{2}+\frac{1}{2})} = 9^1 = 9$, therefore $9^{\frac{1}{2}} = 3 = \sqrt{9}$
Similarly $\sqrt[3]{125} = 125^{\frac{1}{3}}$ and in general $a^{\frac{1}{x}} = \sqrt[x]{a}$
$8^{\frac{2}{3}} = (8^2)^{\frac{1}{3}} = \sqrt[3]{8^2} = (8^{\frac{1}{3}})^2 = (\sqrt[3]{8})^2$
In general $a^{\frac{x}{y}} = \sqrt[y]{a^x} = (\sqrt[y]{a})^x$

***Example** Find the value of (a) $8^{\frac{1}{3}}$ (b) $16^{\frac{3}{4}}$ (c) $125^{\frac{2}{3}}$ (d) $25^{-\frac{3}{2}}$.

(a) $8^{\frac{1}{3}} = \sqrt[3]{8} = \mathbf{2}$, since $2 \times 2 \times 2 = \mathbf{8}$
(b) $16^{\frac{3}{4}} = (\sqrt[4]{16})^3 = 2^3 = \mathbf{8}$
(c) $125^{\frac{2}{3}} = (\sqrt[3]{125})^2 = 5^2 = \mathbf{25}$
(d) $25^{-\frac{3}{2}} = \dfrac{1}{25^{\frac{3}{2}}} = \dfrac{1}{(\sqrt{25})^3} = \dfrac{1}{5^3} = \dfrac{\mathbf{1}}{\mathbf{125}}$

The fractional powers may now be inserted in between the integer powers, as shown in Figure 2 below for the powers of 4.

$4^{-\frac{1}{2}}$	4^0	$4^{\frac{1}{2}}$	4^1	$4^{\frac{3}{2}}$	4^2	$4^{\frac{5}{2}}$
$\frac{1}{2}$	1	2	4	8	16	32

Fig. 2 Fractional and integer powers of 4

SAMPLE QUESTIONS
1 Find the value of: (a) 3^5 (b) 7^3 (c) 10^6 (d) 1^9
2 Express the first number as a power of the second: (a) 64, 2 (b) 81, 3 (c) 64, 8 (d) 100 000, 10 (e) 1024, 2 (f) 625, 5
3 Find calculator approximations (3SF) to: (a) $\sqrt{7}$ (b) $\sqrt{10}$ (c) $\sqrt[3]{25}$ (d) $\sqrt[3]{2}$ (e) $\sqrt{1000}$ (f) $\sqrt[3]{1030}$ (g) $\sqrt{50}$
4 Use a decimal search method to find the following surds accurate to the given accuracy: (a) $\sqrt{5}$, (2DP) (b) $\sqrt{20}$, (3SF) (c) $\sqrt[3]{10}$, (1DP)
5 Use the rules of indices to simplify: (a) $2^7 \times 2^5$ (b) $3^8 \div 3^5$ (c) $(4^5)^2$ (d) $5^2 \times 3^2$ (e) x^5/x^3 (f) $3y^2 \times y^3$
6 Find the following numbers by changing into standard form and using the rules of indices: (a) 0.00024×8000 (b) $5\,600\,000 \times 7\,000\,000$ (c) $360\,000 \div 0.0009$ (d) $0.000048 \div 600\,000$
*7 Find the value of: (a) $27^{\frac{1}{3}}$ (b) $100^{-\frac{1}{2}}$ (c) $243^{\frac{3}{5}}$ (d) 7^{-3} (e) $16^{-\frac{1}{4}}$ (f) $9^{\frac{3}{2}}$ (g) $512^{\frac{2}{3}}$ (h) $1\,000\,000^{-\frac{5}{6}}$ (i) $8^{-\frac{2}{3}}$

NUMBER
1.7 Know your Calculator

When using a calculator to help 'do the sums' in mathematics, it is very important that you *know your calculator well*. The following sections check through the basic ideas but there is no substitute for *regular use* to familiarise you with what your calculator *can* and *cannot do*.

Order of operations
$8 - 2 \times 3$ — Is the result 18 or 2?
The standard convention for calculating arithmetic expressions is:

Calculate the contents of **Brackets** first, then **Powers**, then **Multiplication** and **Division**, and finally **Addition** and **Subtraction**. Work from left to right if priorities are equal.

So $8 - 2 \times 3 = 8 - 6 = 2$.

Now enter this on your calculator to find whether your calculator uses this convention, known as **full algebraic logic**. If the result is 18 then your calculator does not use algebraic logic and *you* will need to enter each calculation in the correct order to make sure your calculator gets the answer right. In the sum above the key sequence needed is:

| 8 | | − | | (| | 2 | | × | | 3 | |) | | = | .

Example Write down the key sequence to calculate,
(a) 3.14×2.6^2, (b) $2 \sin 43°$, (c) $\dfrac{5(1.83 - 0.74)^2}{0.92^3}$.

The key sequences given here assume full algebraic logic. The digits of a number to be entered are not shown as individual keystrokes. The answer is given in square brackets so that you can check on your own calculator.

(a) 3.14 [×] 2.6 [x^2] [=] [21.2264 = 21.2 (3SF)]
(b) 2 [×] 43 [sin] [=] [1.3639967 = 1.364 (4SF)]
(c) 5 [×] [(] 1.83 [−] 0.74 [)] [x^2] [÷] 0.92 [x^y] 3 [=]
    [7.6288578 = 7.63 (3SF)]

Accuracy
When writing down the final answer from a calculator, always round your answer to a *sensible level of accuracy*, bearing in mind the accuracy of the numbers given in the question. It is usual to use the *same* level of accuracy as given in the question, even though the final answer is often *one level less accurate*. (see Error Analysis section, p. 21)

Example Find the area of a circle of radius 4.83 cm.

Use the formula $A = \pi r^2$, and approximate π by 3.14, if your calculator hasn't got a π key. Round the final answer to 3 significant figures corresponding to the accuracy given in the question.

$A = 3.14 \times 4.83^2 = 73.252746$, so the area is **73.3 cm²** (**3SF**).

Calculators often calculate to a greater number of significant figures than they display and the extra digits are either **truncated** (omitted) or **rounded** (up or down). Sometimes the result is shown in the display, but normally the extra rounded digits are not displayed.

Example Calculate $\frac{2}{3}$ of 6, using your calculator.

Key in $\boxed{2}$ $\boxed{\div}$ $\boxed{3}$ $\boxed{\times}$ $\boxed{6}$ $\boxed{=}$. The answer should be 4, since two thirds of 6 is 4, but your calculator may give 3.9999999 because it truncated the intermediate answer to $2 \div 3$ as 0.6666666, rather than rounding it to 0.6666667.

A calculator which truncates may sometimes give peculiar answers which look like recurring decimals and which ought to be rounded up.

Negative numbers

The correct way to enter a negative number is to use the change sign $\boxed{+/-}$ key after entering the number. So -3 is entered as $\boxed{3}$ $\boxed{+/-}$. Your calculator may generate an error code when you attempt to find powers of negative numbers. In this case *you* will have to decide whether the answer is positive or negative and use your calculator to find the power of the corresponding positive number.

Example Calculate (*a*) $(-5 - -3)^2$ (*b*) $f(-3)$, where $f(x) = 2x^2 - 3x - 5$

(*a*) Key in $\boxed{(}$ $\boxed{5}$ $\boxed{+/-}$ $\boxed{-}$ $\boxed{3}$ $\boxed{+/-}$ $\boxed{)}$ $\boxed{x^2}$ $\boxed{=}$ (**4**)

(*b*) $f(-3) = 2(-3)^2 - 3(-3) - 5$. This may be keyed in as

$\boxed{2}$ $\boxed{\times}$ $\boxed{3}$ $\boxed{+/-}$ $\boxed{x^2}$ $\boxed{-}$ $\boxed{3}$ $\boxed{\times}$ $\boxed{3}$ $\boxed{+/-}$ $\boxed{-}$ $\boxed{5}$ $\boxed{=}$ (**22**)

Estimation and checking

Whenever you are about to use a calculator, make a 1 significant figure *estimate* of the answer first to check whether the answer you get from your calculator is reasonable.

Example Make an estimate of the answer to the following:

(*a*) $\dfrac{(5.74 + 6.21)}{3.86^2}$ (*b*) volume of a matchbox 18.6 mm by 63.8 mm by 33.1 mm.

(*a*) $\dfrac{(5.74 + 6.21)}{3.86^2} \approx \dfrac{6+6}{4^2} = \dfrac{12}{16} \approx \mathbf{1}$

21 Know your Calculator

Using a calculator the answer is 0.8020349 = **0.802 (3SF)**.
(b) The volume is found by calculating 18.6 × 63.8 × 33.1 which may be estimated by 20 × 60 × 30 = 36 000 ≈ **40 000**
On a calculator the volume is 39 279.108 = **39 300 mm³ (3SF)**.

Other checks you may use to eliminate errors are:
Repeat the key sequence either in the same order or, where possible, in a different order, and check the answer is the same.
Write down the calculation before entering the key sequence and then, after keying, write down the full calculator display before rounding the answer. Look at this again and check your estimate.
Sometimes it is possible to *reverse the key sequence* from the answer back to one of the initial numbers as a check.

Error analysis

Whenever any measurement is made the result is only as accurate as the particular measuring device allows. Therefore the numbers we use in calculations are only accurate to some given level.

*__Example__ Give upper and lower bounds for the width of a page, given as 21.6 cm (3SF).

The width, W, given as 21.6 cm (3SF), might be any width between 21.55 cm, the **lower bound**, and 21.65 cm, the **upper bound**, and still be rounded to 21.6 cm accurate to 3 significant figures. Using inequality signs we write this as **21.55 cm $\leqslant W <$ 21.65 cm**.

The accuracy level of numbers in questions is not always noted and so we assume that the written digits are the most accurate that can be given. When calculating with rounded numbers the effect of these rounding errors can be calculated and the accuracy of the final answer adjusted.

*__Example__ The length and width of a rectangular field are given to the nearest 10 m as 560 m and 320 m respectively. Find bounds for the area of the field.

In the question the numbers given are accurate to 2 significant figures. The length, L, and the width, W, of the field satisfy the inequalities 555 m $\leqslant L <$ 565 m, 315 m $\leqslant W <$ 325 m, and the area, A, is found by multiplying L and W. Clearly the greatest and smallest values for A will come from multiplying the greatest and smallest values of both L and W.

$$(555 \times 315) \text{ m}^2 \leqslant A < (565 \times 325) \text{ m}^2.$$
$$174\,825 \text{ m}^2 \leqslant A < 183\,625 \text{ m}^2$$
$$\mathbf{170\,000 \text{ m}^2 \text{ (2SF)} < A < 190\,000 \text{ m}^2 \text{ (2SF)}}$$

This gives a *range of answers* accurate to 2 significant figures, but if a *single answer* is required then the best answer that can be given is 200 000 m² (1SF).

Know your Calculator 22

In a question where the numbers are given as accurate to 3 significant figures the final answer can often only be given to 2 significant figures. The only *safe* way is to calculate the upper and lower bounds for the final answer.

SAMPLE QUESTIONS
1. Write down the key sequence for the following calculations, and then use your calculator to check.

   (a) $\dfrac{18}{6+3}$ [2] (b) $3^2 - 5$ [4] (c) 4×3^2 [36]

   (d) $4 \times 7 - 2 \times 6$ [16] (e) $-2 - -3 - 4$ [-3]

   (f) $-5 \times -8 \div -4$ [-10] (g) $\dfrac{-8+3}{-2--3}$ [-5]

   (h) $\dfrac{3(-2)^3}{(-4)^2}$ [-1.5]

2. Check that your calculator gives the correct answer to these calculations: (a) $(3 \div 7) \times 7$, [3] (b) $\tfrac{2}{3} \times \tfrac{3}{5}$, [0.4] (c) $\tfrac{1}{9} + \tfrac{8}{9}$, [1] (d) $\tfrac{3}{7} \div \tfrac{1}{7}$ [3] (e) $\sqrt{81}$ [9] (f) $\sqrt{(3^2 + 4^2)}$ [5]

3. Estimate a 1 significant figure approximation to the following calculation: (a) $(7.93 + 2.14) \times 3.85$ (b) $7.2^2 - 3.1^2$

   (c) $5.78(8.4)^2$ (d) $\dfrac{36.9}{3.21} - \dfrac{11.7}{2.85}$ (e) $(532 + 927 + 1387)/3$

   (f) $\sqrt{(5.74^2 + 8.12^2)}$

4. Write down two different key sequences which might be used to calculate and check the following: (a) $746.8 + 834.9 + 241.7$ (b) $25.6 \times 59.6 \times 38.9$

   (c) $\dfrac{(14.74 + 9.63)}{(8.62 + 5.78)}$ (d) $138.7 - 49.3 - 89.3$ (e) $17.7 \div 3.28 \div 2.59$

*5. Write down upper and lower bounds for the following measurements.
   (a) 17.8 cm (3SF) (b) 0.0072 mm (2SF) (c) 160 g (2SF)
   (d) 35 700 km (3SF) (e) 19.0 m (3SF) (f) 5700 kg (4SF)
   (g) 4.56 cm² (3SF) (h) 5 miles (1SF)

*6. Find upper and lower bounds for:
   (a) the area of a square piece of card of edge 18.3 cm (3SF),
   (b) the volume of a brick with edges 54 mm, 71 mm and 190 mm (2SF),
   (c) the average weight of two boys weighing 37.42 kg and 43.86 kg (4SF).

NUMBER
1.8 Ratio and Proportion

Ratio
A **ratio** is used to compare the **relative sizes of like quantities**.

'There are twice as many girls as boys in this class' becomes 'the ratio of girls to boys is $2:1$'.
'Darren's height is two thirds Robert's height' becomes 'the ratio of Darren's height to Robert's height is $2:3$'.

The numbers in a ratio may be multiplied or divided by any number to produce an **equivalent** ratio. Ratios and fractions share this type of equivalence under multiplication and division.

Example Express as a ratio in its simplest form: (a) the ages of Jane, 6 years old, and Katie, 3 years old. (b) the proportions in a cake recipe requiring 250 g flour, 200 g fat and 150 g sugar. (c) the time spent by two boys on Maths homework, 45 minutes and 1 hour 20 minutes. (d) the length of an aircraft, 40 m, to the length of a scale model, 25 cm.

(a) Jane's age:Katie's age $= 6:3 = \mathbf{2:1}$, (dividing both numbers by 3).
(b) Flour:fat:sugar $= 250:200:150 = \mathbf{5:4:3}$, (dividing by 50).
(c) The two times must be expressed in **common units** first, minutes in this case, so the ratio is $45:(60+20) = 45:80 = \mathbf{9:16}$.
(d) Using metres as the common unit, the ratio of real to scale aircraft is $40:0.25 = \mathbf{160:1}$, (multiplying both numbers by 4 to make them whole).

In cases like part (d) of the last example where the ratio is of the form $n:1$ or $1:n$, the number n is called the **scale** or **scale factor** and this expresses the idea that one quantity is n times the other.

Proportion
Two quantities are in **direct proportion** if there is a constant ratio between them. So if one is halved then so is the other.

Example Rope costs £0.45 per metre. (a) What is the cost of 50 metres of rope? (b) How much rope can be bought for £12?

The length, L, and the cost, C, of the rope are in direct proportion, wth $L:C$ in the ratio $1:0.45$, using the units as given.
(a) To find C when L is 50 we multiply this ratio by 50.
$L:C = 1:0.45 = 50:50 \times 0.45 = 50:22.5$, so the cost is £**22.50**.
(b) To find L when $C = 12$ we first divide by 0.45 and then multiply by 12.
$L:C = \frac{1}{0.45}:1 = \frac{12 \times 1}{0.45}:12 = 26.6:12$, so the length of rope is **26.7 m (3SF)**.

Example There is $\frac{1}{12}$ pint of cream in every pint of full cream milk. (a) How much cream is there in a gallon of full cream milk? (b) How many pints of full cream milk will produce 20 pints of skimmed (no cream) milk?

Ratio and Proportion 24

The quantities of full cream milk, M, and cream, C, are in direct proportion, in the ratio $M:C = 1:\frac{1}{12}$.
(a) We need to find C when $M = 8$ since there are 8 pints in 1 gallon, so multiplying the ratio by 8 we have $M:C = 8:\frac{8}{12}$, and there are $\frac{8}{12} = \frac{2}{3}$ **pint** of cream in 1 gallon of full cream milk.
(b) The quantities of full cream milk, M, and skimmed milk, S, are in direct proportion, in the ratio $M:S = 1:\frac{11}{12}$. We need to find M when $S = 20$, so we need to multiply or divide this ratio until we get ?:20.
$M:S = 1:\frac{11}{12} = 12:11$ (multiplying by 12)
$\qquad = \frac{12}{11}:1$ (dividing by 11)
$\qquad = 20 \times \frac{12}{11}:20$ (multiplying by 20)
$\qquad = 21.8181:20$
So **21.8 pints** (3SF) of full cream milk produces 20 pints of skimmed milk.

Sharing in a given ratio or proportion

Example The winnings of a 3 man syndicate are to be shared in the ratio 1 : 3 : 4. When the prize is £12 000 how much does each person receive?

Since the ratio is 1:3:4 it is convenient to count these as a total number, $1 + 3 + 4 \ (= 8)$, of equal shares. Dividing the £12 000 by 8 makes each share worth £1500. 3 shares and 4 shares are worth $3 \times £1500$ and $4 \times £1500$ respectively and so the three people will receive **£1500**, **£4500** and **£6000** respectively.

Example Calculate the angles needed to draw a piechart to represent the data on car colours: red 45, blue 38, green 17, gold 13, black 7 cars.

The problem is to share out a full turn, 360°, in the same proportions as the colours of the cars. The total number of cars is $45 + 38 + 17 + 13 + 7 = 120$, so each car is worth $360/120 = 3°$.
The angles for each sector may be found by multiplying the number of cars by 3. The results are shown in the table opposite which can also be used to check the working.

⟵ × 3 ⟶

Colour	Number	Angle
Red	45	135
Blue	38	114
Green	17	51
Gold	13	39
Black	7	21
Total	120	360

Inverse proportion

The relationship between some quantities is such that if you *increase* one quantity you must *decrease* the other.

If the speed is *doubled* over a set distance then the time taken is *halved*.

If the number of workers is *trebled* (multiplied by 3) then the time taken to finish the job is reduced to *one third* (divided by 3).

Such quantities are said to be in **inverse proportion**.

25 Ratio and Proportion

Example A car is being tested for fuel consumption by running it around a track at different speeds for a set distance. The time taken is inversely proportional to the speed. On one run lasting 45 minutes the car travelled at a speed of 40 mile/h.
(a) How long will a run last at a speed of 20 mile/h?
(b) At what speed will the car produce a run lasting 1 hour?

(a) At a speed of 40 mile/h the run lasts 45 minutes.

$\qquad\qquad\qquad\quad \div 2 \qquad\qquad\qquad \times 2$

So at a speed of 20 mile/h the run lasts $45 \times 2 =$ **90 minutes**, since if the speed is *halved* then the time is *doubled*.

(b) A run lasts 45 minutes at a speed of 40 mile/h.

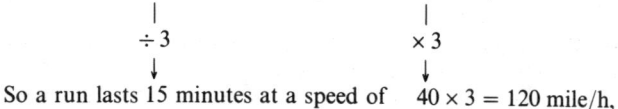

$\qquad\qquad\quad \div 3 \qquad\qquad\qquad\qquad \times 3$

So a run lasts 15 minutes at a speed of $40 \times 3 = 120$ mile/h,

$\qquad\qquad\quad \times 4 \qquad\qquad\qquad\qquad \div 4$

and a run lasts 1 hour (60 min) at a speed of $120 \div 4 =$ **30 mile/h**.

Example The number of plastic widgits manufactured in one production run is inversely proportional to the unit cost and directly proportional to the time taken. If it costs £300 for a production run of 10 000 widgits lasting 6 hours what is the unit cost. How many will need to be manufactured to bring the unit cost down to 2p? How long will a production run of 8000 widgits take?

The total cost of 10 000 widgits is £300, so the unit cost on a run of 10 000 widgits is £300 ÷ 10 000 = **3p**.

The unit cost is 3p for a run of 10 000 widgits, so

$\qquad\qquad\quad \div 3 \qquad\qquad\quad \times 3$

the unit cost is 1p for a run of $10\,000 \times 3 = 30\,000$ widgits, and

$\qquad\qquad\quad \times 2 \qquad\qquad\quad \div 2$

the unit cost is 2p for a run of $30\,000 \div 2 =$ **15 000 widgits**.

For a run of 10 000 widgits the time taken is 6 hours, so

$\qquad\qquad\quad \times \tfrac{4}{5} \qquad\qquad\quad \times \tfrac{4}{5}$

for a run of 8000 widgits the time taken is $6 \times \tfrac{4}{5} = 4.8$ hours, and 4.8 hours = 4 h (0.8 × 60) min = **4 hours 48 minutes**.

Ratio and Proportion 26

Note: In everyday usage, speeds are most commonly expressed in terms of miles per hour (mph *or* mile/h) and
kilometres per hour (kph *or* km/h).

To help you familiarise yourself with the alternative forms of these abbreviations, all have been used in this book. In GCSE work and examinations, you should always use the same notation in your answer as is used in the question.

SAMPLE QUESTIONS

1. Express the following ratios in simplest form:
   (a) 3 cm:24 mm, (b) 15p:£1.80, (c) 250 g:3.5 kg,
   (d) 3 days:6 weeks.
2. Give the scale factor for the following ratios by expressing in the form $1:n$ (a) 50 cm:2 km, (b) 3 ounces:3 lb,
   (c) 10 s:1 day, (d) 2.5×10^6 m:5×10^9 m, (e) 5p:17p,
   (f) 0.8 mm:75 cm.
3. The ratio between the amount of oil and vinegar needed in a French dressing is 5:2. (a) How much oil must be added to 300 ml of vinegar? (b) How much oil and vinegar is needed to make 1 litre of French dressing?
4. A rope stretches 2 cm when pulled by a force of 300 Newtons. (a) How much will it stretch when pulled by a force of 500 Newtons? (b) What force will stretch it by 4.5 cm?
5. Picture framing costs £1.25 per metre. What is the cost for a frame 30 cm by 55 cm (assume you use wood equal in length to the perimeter of a rectangle with these dimensions). What is the widest picture frame of height 25 cm that could be made from framing costing £5.00?
6. The proceeds of a Council lottery are shared between 3 charities in the ratio 5:2:1. The proceeds total £24 000. How much does each charity receive?
7. Calculate the angles needed to draw a piechart showing the proportions of votes cast in an election with the following poll:
   Democrats 24 750, Republicans 35 520, Independents 11 730.
8. The mass of 1000 litres of heating oil is 850 kg. (a) What is the maximum mass of heating oil in a 900 l domestic tank. (b) What is the maximum capacity for a tanker which has a load limit of 30 tonne?
9. It takes 2 men 20 hours to build a brick wall. (a) How many men would be needed to cut the time taken to 5 hours? (b) How long would it take 3 men to build the same wall?

NUMBER
1.9 Percentages

A **percentage** is a fraction expressed in hundredths, 'per cent' means 'out of 100'.

$25\% = \frac{25}{100} = 0.25$, since $25 \div 100 = 0.25$

When money is used it is easier to think of 12% as **12 pence in the pound**.

Calculations with percentages often involve changing a percentage into a fraction or decimal number and changing a decimal back into a percentage.

Example Convert the number given into its two remaining forms, from percentage, fraction and decimal: (a) 85%, (b) $8\frac{1}{2}\%$, (c) 0.375, (d) $\frac{1}{3}$.

(a) $85\% = \frac{85}{100} = \mathbf{0.85}$ (85p in the pound).

(b) $8\frac{1}{2}\% = \dfrac{8.5}{100} = \frac{85}{1000} = \mathbf{0.085}$ (note the zero in the tenths column).

(c) $0.375 = \frac{375}{1000} = \dfrac{37.5}{100} = \mathbf{37\frac{1}{2}\%}$

(d) $\frac{1}{3} = \mathbf{0.33\dot{3}} = \mathbf{33\frac{1}{3}\%}$

Example Mr Jones invests £2500 and is paid 15% at the end of the first year as a dividend. (a) How much is this dividend? The following year he is paid a dividend of £450. (b) What percentage dividend is this?

This problem may be solved several ways, and we show two methods here.

Calculator method using **decimals**:
(a) $15\% = 0.15$. So 15% of £2500 $= 0.15 \times 2500 = \mathbf{£375}$, using a calculator to do the multiplication.
(b) £450 is $\frac{450}{2500}$ of £2500, expressed as a fraction.
$\frac{450}{2500} = 0.18 = \mathbf{18\%}$ as a percentage, carrying out the division of 450 by 2500 directly on the calculator.

Fraction method which may be used when **no calculator** is available:
(a) $15\% = \frac{15}{100}$
15% of £2500 $= \frac{15}{100}$ of £2500 $= \frac{15}{100} \times 2500 = 15 \times 25 = \mathbf{£375}$
(b) £450 is $\frac{450}{2500}$ of £2500, expressed as a fraction.
$\frac{450}{2500} = \frac{18}{100} = 0.18 = \mathbf{18\%}$, converting the fraction into a percentage.

Percentage increase and decrease

Increasing or decreasing an amount by a certain percentage may be calculated by finding the increase or decrease and then adding or subtracting. The following **scale factor method** is rather easier and very suitable for calculator use.

Percentages 28

Example (a) Increase £24 by 15%, (b) decrease £2.45 by 30%.

(a) 15% of £24 = 0.15 × 24
 + original £24 = 1   × 24
Increase £24 by 15% = 1.15 × 24
   = **£27.60**

(b) original £2.45 = 1   × 2.45
 − 30% of £2.45 = 0.30 × 2.45
Decrease £2.45 by 30%1 = 0.70 × 2.45
   = **£1.72**

Increasing by 15% is equivalent to *multiplying by* 1.15.

$$\boxed{\text{Original Price}} \xrightarrow{\times 1.15} \boxed{\text{Price} + 15\%}$$

Decreasing by 30% is equivalent to *multiplying by* 0.70, (finding 70%).

$$\boxed{\text{Original price}} \xrightarrow{\times 0.70} \boxed{\text{Price} - 30\%}$$

Example Find the final answer after the following changes.
(a) Increase £100 by 12%, (b) Decrease 250 km by 10%,
(c) Decrease 3 kg by 55%, (d) Increase £9500 by 7.8%.

(a) Increase £100 by 12% = £100 × 1.12 = **£112**
(b) Decrease 250 km by 10% = 250 km × (1 − 0.10)
     = 250 km × 0.90 = **225 km**
(c) Decrease 3 kg by 55% = Find 45% of 3 kg = 3 kg × 0.45 = **1.35 kg**
(d) Increase £9500 by 7.8% = £9500 × 1.078 = **£10 241**

Finding the percentage change

This scale factor idea may now be used, in reverse, to find the percentage by which an original amount has been increased or decreased.

Example Find the percentage change when:
(a) a man's annual salary is increased from £8650 to £9342,
(b) a coat is reduced in a sale from £59 to £49.

(a) $\boxed{\text{Original Salary}} \xrightarrow{\times \text{scale factor}} \boxed{\text{New Salary}}$

To find the scale factor we divide the new salary by the original salary.
Scale factor = 9342 ÷ 8650 = 1.08, and so his salary has been increased by **8%**.

Percentages

(b)

```
Original Price  ──× scale factor──▶  Sale Price
```

To find the scale factor we divide the sale price by the original price. Scale factor = $49 \div 59 = 0.83$ (2DP) $= 1 - 0.17$ and so the percentage reduction is **17%**.

Example A house first valued in 1985, was revalued 15% higher in 1986 and 10% higher again in 1987. What is the percentage increase in the value of the house over the two years?

```
1985 value  ──× 1.15──▶  1986 value  ──× 1.10──▶  1987 value
            ──────────── × scale factor ────────────▶
```

The original 1985 value has been mutiplied by 1.15 and 1.10 to get the 1987 value. So the scale factor is $1.15 \times 1.10 = 1.265$, and the house has increased in value by $26\frac{1}{2}$ % over the two years.

Notice that the increase is more than the sum of the two percentages since the 10% increase is taken on the 1986 value rather than the 1985 value.

Finding the original price

Finally we again use the scale factor method to solve the problem of finding the original price given the final price and percentage increase.

Example After a general 5% increase in new car prices a car is priced at £6720. What was its price before the increase?

```
Original Price  ⇌ × 1.05 / ÷ 1.05 ⇌  New Price
```

To increase the original price to the new price we *multiplied* by the scale factor 1.05, so we *divide* the new price by 1.05 to find the original price. Original Price = £$6720 \div 1.05 =$ **£6400**.

Note: Decreasing £6720 by 5% does not give the same answer, since the 5% increase was 5% of £6400, not 5% of £6720.

Example A stereo radio/cassette player is advertised as 30% OFF   ONLY £49. What was the price before the reduction?

```
Original Price  ⇌ × 0.70 / ÷ 0.70 ⇌  Reduced Price
```

To reduce the price by 30% the original price has been *multiplied* by the scale factor 0.70 ($= 1 - 0.30$). So to find the original price we *divide* the reduced price by 0.70. Original Price = £$49 \div 0.70 =$ **£70**. This answer may be checked by reducing £70 by 30% which gives £49.

*Repeated percentage increases

*__Example__ The area covered by a bacterial growth spreads at the rate of 18% every 12 hours. If its initial area is 5 mm², what will be the area after 3 days? After how many 12 hour intervals will the growth exceed 1 cm²?

Percentages 30

$$\boxed{\text{Initial area}} \xrightarrow{\times 1.18} \boxed{\text{12 hours later}} \xrightarrow{\times 1.18} \boxed{\text{24 hours later}} \xrightarrow{\times 1.18} \boxed{\text{36 hours later}}$$

After 3 days (6 × 12 h) the bacteria's area will be multipled by 1.18 six times.
Final area = 5 mm² × 1.18 × 1.18 × 1.18 × 1.18 × 1.18 × 1.18
 = 5 mm² × 1.18^6
 = **13.5 mm²** (3SF)
1 cm² = 10 mm × 10 mm = 100 mm².
Assuming that n 12-hour intervals are needed, we have to solve the inequality
$$5 \times 1.18^n \geqslant 100.$$
Dividing both sides by 5 $1.18^n \geqslant 20$
By repeated trials on a calculator we find $1.18^{18} = 19.7$ (3SF)
and $1.18^{19} = 23.2$ (3SF).
Therefore it takes **19** 12-hour intervals before the area exceeds 1 cm².

SAMPLE QUESTIONS
1 Express the following fraction as a decimal and a percentage:
 (a) $\frac{1}{20}$ (b) $\frac{7}{10}$ (c) $\frac{4}{5}$ (d) $\frac{3}{8}$ (e) $\frac{2}{3}$ (f) $\frac{5}{12}$ (g) $\frac{3}{2}$
2 Express the following percentage as a decimal and a fraction:
 (a) 20% (b) 75% (c) 6½% (d) 88% (e) 106% (f) 12½%
3 Find: (a) 15% of 360 km   (b) 7% of £9.50
 (c) 52% of 38 465 votes   (d) 2½% of £7580   (e) 120% of 3500 ants
 (f) 0.5% of 3.4 mm
4 Find the new quantity when:
 (a) the population of a village (256) is reduced by 16%,
 (b) a tent, weighing 35 kg, has its weight increased by 85% due to rain,
 (c) VAT at 15% is added to the price of a television costing £349,
 (d) cinema admissions (125 million per year) fall by 28%.
5 Find the percentage change when:
 (a) a child's weight increases from 32 kg to 38 kg,
 (b) the production of tea bags decreases from 524 million to 478 million,
 (c) the number of pupils attending Brainwood School rises from 965 to 1045.
6 Find the original quantity when:
 (a) a slimmer's weight is down 12% and she now weighs 65 kg,
 (b) the area of a forest has been increased by 8% and is now 256 km²,
 (c) the price of a Lazaratti car has increased by 35% to £18 500.
*7 The height of a young tree, 0.8 m when planted, increases by 12%, 8% and 10% in the first three years. Find the height after 3 years and the total percentage increase.
*8 A colony of rare Sea Sparrows is estimated to be declining by 5% per year. The colony numbered 700 in Summer 1983. How large will the colony be in Summer 1988 and when it will first fall below 300?

EVERYDAY ARITHMETIC
2.1 Buying and Selling

Profit and loss

When an article is bought at one price and then sold at a higher price the difference between the selling price and the buying price is called the **profit**. When the selling price is less than the buying price the difference is known as the **loss**.

Example Find a garage's profit or loss on the following:
(a) an Adagio bought for £1500 and sold for £1650,
(b) a Centtro bought for £3200 and sold for £3680,
(c) a Buletta bought for £750 and sold for £570.

(a) £1650 > £1500 so the profit is £1650 − £1500 = **£150**.
(b) £3680 > £3200 so the profit is £3680 − £3200 = **£480**.
(c) £570 < £750 so the loss is £750 − £570 = **£180**.
The overall profit is £150 + £480 − £180 = **£450**.

Percentage profit or mark-up

To obtain a fixed **percentage profit** the selling price of an article is calculated by increasing the price paid by this percentage mark-up.

Example A clothes retailer marks up all goods by 25%. Find the selling price of vests bought at £2, shirts at £5.80 and ties at 88p. Calculate the profit on selling 5 shirts, 3 vests and 2 ties.

Increasing by 25% can be found by multiplying by 1.25 (page 28).
The selling price of a vest is £2 × 1.25 = **£2.50**.
The selling price of a shirt is £5.80 × 1.25 = **£7.25**.
The selling price of a tie is 88p × 1.25 = **£1.10**.
The total cost, to the retailer, of 5 shirts, 3 vests and 2 ties is
$(5 \times £5.80) + (3 \times £2) + (2 \times £0.88) =$ **£36.76**.
The profit will be 25% of £36.76 = £36.76 × 0.25 = **£9.19**.

Example Mr Singh had his car serviced. His bill consisted of £23.40 for spare parts and £33.60 for labour, both exclusive of VAT. Calculate: (a) VAT on his bill at 15% (b) the total cost including VAT.

(a) VAT at 15% is 0.15 × (£23.40 + £33.60) = 0.15 × £57 = **£8.55**.
(b) Total inclusive charge is £57 + £8.55 which is **£65.55**.

Family budget

A typical family budget can be split into two headings:
Income—money *coming in* from wages, child benefits and other sources.
Outgoings—money *spent* on rent, rates, food and household goods.
The difficulty comes in satisfying the inequality

Outgoing ≤ Income .

Buying and Selling 32

Example The Jones' weekly budget includes net pay £74.50, rent £16.80, rates £6.70, electricity, gas and water charges £12, child benefits £13.70. They normally spend £35 on food and household goods. How much spare money will the Jones family normally have each week?

Income: pay and child benefits £74.50 + £13.70 = £88.20.
Outgoing: rent, rates, charges and food
£16.80 + £6.70 + £12 + £35 = £70.50.
Spare money each week is   £88.20 − £70.50 = **£17.70**.

SAMPLE QUESTIONS

1. Find the total cost of 2 pots of honey at 69 p per pot, 8 tins of soup at 47 p per tin, 500 g of drinking chocolate if 250 g cost 41 p, and 3 kg of sugar at 46 p per kilo.   £7·34

2. A man goes into a hardware shop and buys a hammer costing £3.26, netting costing £4.27, and 5 packets of nails at 39 p each. Calculate his final bill if VAT at 15% must be added to the total cost.

3. A firm of printers charges £15 setting up fee and 5.6 p per poster to print a poster. Calculate the cost of printing 500 posters if VAT at 15% must be added to the given costs.

4. The price of a table and six chairs is reduced in a sale by 20%. What is the sale price if the original price is £325?

5. Calculate the selling price if the mark-up is 25%: (*a*) radio £33, (*b*) turntable £57.40, (*c*) amplifier £40.40, (*d*) speakers £59 a pair.

6. Tom bought a bicycle for £25 and sold it to Dick at a profit of 20%. Dick later sold it to Harry at a loss of 15%. Calculate what Dick and Harry paid for the bicycle.

7. The price of a scooter, originally £185.90, had to be increased by 12% due to rising production costs. What should be its price in a sale where all goods are offered at 10% off?

8. Seven tickets cost £8.75. What is the cost of: (*a*) 1 ticket, (*b*) 3 tickets?

9. What is the difference in total price between 2 books costing £1.89 each and 3 books costing £1.49 each?

10. Ms Nelson's net weekly pay is £92.40. She pays out on rent £18.60, rates £7.45, and electricity £4.50 each week. How much surplus money does she have this week after spending £28.49 on food and £17.50 on new clothes.

11. Frozen beans cost 38 p per 500 g and fresh beans cost 60 p per kg. Which is cheaper and by how much if 3 kg are required?

12. A calculator, normally priced at £19.82, is offered at £15.45 for a limited period. What percentage discount is being offered and how much is saved if 6 calculators are purhased during the offer?

EVERYDAY ARITHMETIC
2.2 Saving and Borrowing

2

When a sum of money, **the principal**, is deposited as **savings** in a bank or building society it earns **interest** at the stated **rate of interest**. The rate is usually given as a **annual percentage rate** (APR), although the interest may be calculated half yearly or more frequently. When a sum of money is borrowed for a period of time the borrower will have to pay interest as well as the principal when the loan is repaid.

Income or growth
The interest on savings may be taken as **income**. The principal remains steady and the same interest, **simple interest**, is paid each year. Alternatively the interest may be added to the principal, **compounded**, and then the principal and the amount of interest increase each year.

Example received, On a deposit of £580 at 8.75% for 3 years, compare: (a) the income (b) the growth under compound interest.

Income: (a) Simple interest each year is
8.75% of £580 = 0.0875 × £580 = £50.75.
Over 3 years the total income will be 3 × £50.75 = **£152.25**.

Growth: (b) Use the scale factor method, page 28, to increase by 8.75%.
After 1 year the principal will be £580 × 1.0875 = £630.75
After 2 years the principal will be
£630.75 × 1.0875 = £685.94
After 3 years the principal will be £685.94 × 1.0875 = £745.96
(or alternatively after 3 years the principal = £580 × 1.0875^3)
The principal has grown by £745.96 − £580 = **£165.96**.

Borrowing
Hire Purchase, or **HP**, is a method of borrowing money to purchase household goods. A percentage of the total price is paid at purchase as a **down payment** or **deposit**, and regular monthly **repayments** are made for a fixed **period**.

Interest is charged on the amount of the loan **still outstanding** and the monthly repayments are calculated so that the loan is paid off by the last repayment.

Example Calculate the extra paid in purchasing a cooker priced at £359 by an HP scheme, deposit 15% and 24 monthly payments of £15.25.

The deposit is 15% of £359 = 0.15 × £359 = £53.85.
The 24 payments total 24 × £15.25 = £366.00.
Total HP cost is £53.85 + £366.00 = £419.85.
Extra paid is HP cost − Cash cost = £419.85 − £359 = **£60.85**.

Saving and Borrowing 34

A Repayment Mortgage is a method of borrowing money to buy or improve a house and works in the same way as Hire Purchase. A **95% mortgage over 25 years** means that **5%** of the purchase price of the house must be paid as **deposit** and the remainder **95%** is loaned by the building society or bank to be repaid by monthly **repayments** over the next 25 years.

Example A young couple decide to buy a house costing £30 000. They are offered a 90% mortgage over 25 years, with monthly repayments of £215. Calculate the amount of deposit they must pay initially and the total amount invested in buying the house before they own it all.

The deposit on a 90% mortgage is
10% of £30 000 = $0.10 \times £30\,000 = £3000$. The total monthly payments will be $£215 \times 12 \times 25 = £64\,500$, so the total amount invested in buying the house is $£3000 + £64\,500 = £67\,500$.

Example Recess credit card company charge 2% on any amount outstanding each month. In May Mrs Andrews receives notice that she owes Recess £184.36 and she decides to pay £50. What will she owe in June if she makes no further purchases with her Recess card?

After her payment of £50 has been received by Recess her account will stand at $£184.36 - £50 = £134.36$.
Increasing this balance by 2% gives $1.02 \times £134.36 = £137.05$ (2DP), and her account will stand at **£137.05** in June.

SAMPLE QUESTIONS
1. Calculate the annual interest earned on (a) £225 at 12% (b) £58.60 at 7.8% (c) £38 500 at 14.3% (d) £2540 at $11\frac{1}{4}$%
2. Paul is considering whether to invest £1500 for income or growth. At 9.5% for 2 years calculate: (a) the total income, (b) the growth.
3. Miss Young bought a car costing £6500 on HP. She paid a 20% deposit and 36 payments of £186.40. What was the total HP cost?
4. Elisabeth buys a TV set on HP, paying £50 deposit and 12 monthly payments of £22.50. What is the total HP cost, and how much less would the TV have cost if she could have paid the cash price of £285?
5. Mr and Mrs Grant are offered an 85% mortgage on a house costing £28 500. How much deposit must they pay? The repayments are £215.35 a month for 20 years. How much have they spent altogether in purchasing the house?
6. Mr Grand's deposits at the bank rose from £13 560 at the start to £14 848.20 by the end of the year. What rate of interest is he receiving?
*7 Mr Fall takes out a Life Policy at £23.16 per month. The policy pays out £25 000 if he dies during the next 20 years. How much does he pay in total if he survives the full 20 years. If he dies after 15 years what percentage of the sum assured has he paid?

EVERYDAY ARITHMETIC
2.3 Earning a Wage

Rates of pay
Employees are normally paid either each week or each month. The **gross pay** of **weekly paid** employees is often calculated at **basic rate** per hour for a **basic working week**, usually 40 hours. When extra hours are worked then **overtime** at a higher rate is paid. **Monthly paid** employees are paid an **annual salary**, and their gross monthly pay is calculated by dividing by 12. Sales staff are sometimes paid a basic wage plus **commission**, a percentage of the value of the goods or services paid. Production workers may be paid at **piece rate**, an amount for each article or job completed, or paid a **bonus** for increased productivity.

Example Mrs Smith works for 40 hours basic at £2.35 an hour, plus overtime at 'time and a half'. Calculate her gross pay in a week when she works $2\frac{1}{2}$ hours overtime on Wednesday and 5 hours on Saturday.

The overtime rate 'time and a half' means $1\frac{1}{2} \times £2.35 = £3.53$ per hour.
Mrs Smith's basic wage is $40 \times £2.35 = £94.00$.
Her overtime pay is $(2\frac{1}{2} + 5) \times £3.53 = 7\frac{1}{2} \times £3.53 = £26.44$.
So her total gross pay is $£94.00 + £26.44 = \pmb{£120.44}$.

Example Mr Macdonald is paid a basic annual salary of £6870 plus 10% commission on any sales above £800 per month. Find his gross pay in an exceptional month where he sells £3640 worth of goods.

Mr Macdonald earns commission this month on $£3640 - £800 = £2840$,
so his commission is 10% of $£2840 = 0.10 \times £2840 = £284$.
His basic monthly pay is $£6870 \div 12 = £572.50$,
so his gross pay this month is $£572.50 + £284.00 = \pmb{£856.50}$.

Tax and deductions
Every wage earner pays **Income tax** to the Government for public services, defence, etc. A **tax free allowance** is allocated depending on personal circumstances, and the remainder of gross pay, known as **taxable pay**, is taxed at a basic rate of 30%, subject to changes in taxation policy. Higher rates of tax 40%, 45%, are charged if one's taxable pay is above certain limits.

> Gross pay − tax free allowance = taxable pay

Other **deductions** include, **National Insurance**, for pension and sickness benefits, **superannuation**, for an increased pension, and **other deductions**, for special clothing, union dues, or fees.

> Gross pay − Deductions = Net pay

Earning a Wage 36

Example Mr Brown has a gross monthly salary of £1037.80, a married man's allowance of £26.50. He pays £56.93 National Insurance per month, 6% of gross superannuation and basic rate Income tax at 30%. Calculate his monthly tax, deductions and net pay to fill out his monthly payslip.

Gross pay	1037.80	Tax 30% × 777.30 = 233.19	Gross pay	1037.80
Tax allowance	260.50	National Insurance = 56.93	Deductions	352.39
		Sup. 6% × 1037.80 = 62.27		
Taxable pay	777.30	Total deductions = 352.39	Net pay	685.41

SAMPLE QUESTIONS

1. Mr Johnson is paid £2.20 per hour basic, plus overtime at 'time and a half' for weekdays and Saturdays, and 'double time' on Sundays. Find Mr Johnson's gross weekly pay for: (a) a basic 42 hour week, (b) a basic week plus 3½ hours Friday night and 2 hours Saturday, (c) a basic week plus 5 hours on Saturday and 4½ hours on Sunday.

2. Anne, Brenda, Chris and David work in different shops and earn £1.58, £1.82, £1.16 and £1.76 an hour respectively. Calculate the gross pay when: (a) Anne works 37½ hours basic, (b) Brenda works 45 hours basic plus 3 hours overtime at 'double time', (c) Chris works 36 hours basic and earns £38.27 on commission, (d) David works 42 hours basic, 2 hours at 'time and a half', and 4 hours 'double time'.

3. Calculate the net monthly pay of the following people if they pay basic rate Income tax at 30%:
   (a) Mrs Sultani, annual salary £6576, tax allowance £1375, N.I. £378.50,
   (b) Mr Dunn, gross £9624, Super. £583.44, N.I. £558, tax allowance £2005,
   (c) Mr Biggs, gross £14 624, Super. 7% of gross, N.I. £769, tax allowance £3115.

4. Alan is paid £156.24 gross per week. Find his annual pay (52 weeks).

5. Mrs Fortune has an annual taxable income of £18 462. She pays 30% on the first £15 000, 40% on the next £3000 and 45% on the remainder. How much tax does she pay each month?

6. Isobel's gross wages amount to £4570 in one year. She pays £234 National Insurance, £86 on protective clothing supplied by her firm and has a tax allowance of £2556. What is her net annual pay and how much is this on average each week?

7. Miss Dahl is paid a basic £1.15 an hour, plus piece rate of 24p for each electronic board she completes. Calculate her gross pay last week when she worked a total of 48 hours and completed 288 boards. How long did each board take to finish on average? Find her total deductions if her net pay came to £86.36.

EVERYDAY ARITHMETIC
2.4 Reading Tables

Example Table 1 shows the dimensions of International Paper sizes.
(a) What are the dimensions of A4 paper in millimetres?
(b) Calculate the area of A1, A2 and A3 paper to the nearest square inch.
(c) Express the relative areas of A1, A2 and A3 paper as a ratio.
(d) Into how many sheets of A5 can a single sheet of A0 paper be cut?

(a) The A4 row and the millimetres column give the size **210 × 297 mm**.
(b) The area of A1 paper is 23.39 × 33.11 sq in = 774.44 sq in = **774 sq in**.
   Similarly:
   A2 paper has area 16.54 × 23.39 sq in = 386.87 sq in = **387 sq in**, and
   A3 paper has area 11.69 × 16.54 sq in = 193.35 sq in = **193 sq in**.
(c) The ratio of areas of A1, A2, A3 paper is 774:387:193 ≈ **4:2:1**.
(d) From the given dimensions, A1 paper can be cut into 2 sheets of A2, so
   1 sheet of A0 = 2 sheets of A1 = 4 sheets of A2 = ... = **32** sheets of A5.

International Paper Sizes

Designation	*size*	
	mm	in
A0	841 × 1189	33.11 × 46.81
A1	594 × 841	23.39 × 33.11
A2	420 × 594	16.54 × 23.39
A3	297 × 420	11.69 × 16.54
A4	210 × 297	8.27 × 11.69
A5	148 × 210	5.83 × 8.27

Table 1

British Public Services

Expenditure	(*costs in £millions*)	
	1960	1970
Defence	1612	2461
Roads	172	335
Housing	490	1213
Education	917	2592
Social Services	1488	3908
Total	9400	21564

Table 2

Example Table 2 shows the amount spent on five particular Services, and the total expenditure on all British Public Services in 1960 and 1970. (a) How much more was spent on Defence in 1970 than 1960? (b) Calculate the percentage increase in expenditure on Education between 1960 and 1970. (c) Calculate the percentage increase in total expenditure between 1960 and 1970. (d) Which of the services shown had proportionally less spent on them in 1970 than in 1960?

(a) The Defence row shows the expenditure to have risen from £1612 million to £2461 million, an increase of 2461 − 1612 = **£849 million**.
(b) Education expenditure increased from £917 million to £2592 million, 2592 ÷ 917 = 2.826 = 1 + 1.826, an increase of **183%**.
(c) Total expenditure rose from £9400 million to £21564 million, 21564 ÷ 9400 = 2.294, an increase of **129%**.
(d) Percentage increases for each service are: Defence 53%, Roads 95%, Housing 148%, Education 183%, Social Security 163%, Total 129%. So the services on which proportionally less has been spent in 1970 are **Defence** and **Roads**.

Example Table 3 shows sunrise and sunset times in London in 1976. (a) At what time did the sun set on July 4th? (b) How many hours and minutes of daylight were there on April 4th? (c) What is the percentage increase in hours of daylight between January 4th and July 4th?

(a) On July 4th the sun set at 2120 hrs, **9.20 pm**.
(b) Daylight lasted from 0629 hrs to 1939 hrs, which is **13 h 10 min**.
(c) 0806 to 1605 is 7 h 59 min, which is 479 min.
    0450 to 2120 is 16 h 30 min, which is 990 min, 511 min more.
    The percentage increase in daylight = $511/479 = 1.07(3SF) =$ **107 %**.

Sunrise and Sunset for London 1976

		Rise	Set
January	4	0806	1605
April	4	0629	1939
July	4	0450	2120
October	4	0705	1834

Table 3

Metric Equivalents

Length	1 ft = 30.48 cm
Area	1 sq yd = 0.836 sq m
Capacity	1 gal = 4.546 l
Weight	1 lb = 0.454 kg
Speed	1 mph = 1.609 kph

Table 4

SAMPLE QUESTIONS

1 Using Table 1: (a) What are the dimensions of A1 size paper in centimetres? (b) What is the perimeter of a sheet of A5 paper in inches? (c) A sheet of A1 paper is folded in half three times and is then the same size as another standard size. Which standard size is this?
2 Using Table 2, how much more was spent on Education than Housing in: (a) 1960, (b) 1970?
3 Using Table 3, calculate: (a) what fraction of July 4th is daylight, (b) the difference in hours of daylight between April 4th and October 4th.
4 Using the information given in Table 4, express:
   (a) 3 feet in centimetres,   (b) 100 gallons in litres,
   (c) 1 kilogram in pounds (lb),   (d) 50 kph in mph.
5 Using Table 4, which is greater:
   (a) 100 sq yd or 100 m^2 (b) 200 gal or 900 l   (c) 5 kg or 10 lb
   (d) 2 m or 7 ft   (e) 50 mph or 80 kph   (f) 0.2 gal or 1 l.

EVERYDAY ARITHMETIC
2.5 Time

12 and 24 hour clock

There are in common use two systems of recording time. Figure 1 takes four times during the day, and demonstrates how time is spoken, recorded in the 24 and 12 hour systems, and shown on a clock with hands.

	1 min past midnight	12 min to noon	Half past three in afternoon	Quarter to ten (night)
24 hour clock	0001	1148	1530	2145
12 hour clock	12.01 **am**	11.48 am	3.30 **pm**	9.45 pm

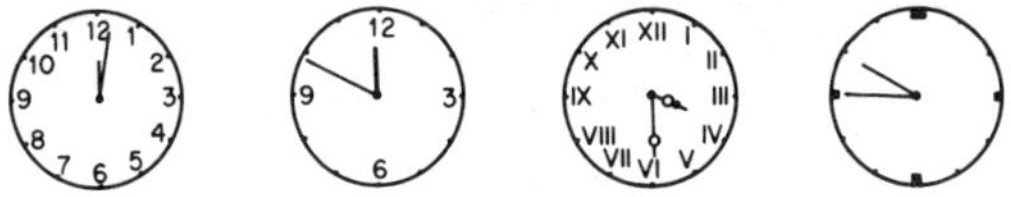

Fig. 1 Different ways of recording time.

Example Natascha gets up at three minutes to seven in the morning and goes to bed again at twenty past nine in the evening. Write these times using the 24 and 12 hours clock systems and calculate how many hours and minutes Natascha was up out of bed.

There are **60 minutes in 1 hour**, so 'three minutes to seven' is 57 minutes past 6 o'clock, written **0657 hrs** or **6.57 am** ('hrs' may be omitted).

There are 12 hours in the morning (am) so 9 o'clock in the evening is 2100 hrs in 24 hours clock time $(12 + 9 = 21)$. 'Twenty past nine' is therefore written as **2120** or **9.20 pm**.

To calculate the time elapsed the temptation is simply to subtract the two 24 hour clock times, $2120 - 0657 = 1463$ which is 40 minutes too much due to there being 60 minutes in an hour, not 100. The method below is easier to understand and can be done in both 24 and 12 hour clock time.

0657 to 0700	0700 to 2100	2100 to 2120	0657 to 2120
3 minutes +	14 hours +	20 minutes =	**14 h 23 min**

Timetables

Example Use the Brightlington–Poolsberry railway timetable to answer: (a) Arriving at Brightlington station at 4.38 pm, which is the earliest train to Poolsberry, when will it get there and how long will the journey take? (b) Which train can I catch from Brightlington to arrive at Faringsham by 3 o'clock in the afternoon? (c) I arrive at Poolsberry station at 1038 to meet my aunt off the 0945 from Brightlington. How long must I wait if the train is announced as running 25 minutes late?

Brightlington	d	0833	0945	1054	1235	1325	1427	1533	1710	1933	2145
Upper Gayford	d	0854	1004	—	—	1347	—	1554	1732	—	2204
Midlinghaven	d	0912	1022	—	—	1400	—	1612	1750	—	2222
Faringsham	d	0925	1035	1122	1304	1416	—	1625	1804	—	2235
Poolsberry	d	0938	1047	1135	1316	1432	1504	1638	1818	2012	2247

(a) 4.38 pm = 1638, so the first train is at **1710**, arriving at Poolsberry at **1818**, taking **1 hour 8 minutes** for the journey.

(b) To arrive at 1500, the **1325** is the latest possible train since the 1427 doesn't stop at Faringsham.

(c) The 0945 normally arrives at 1047, but should arrive at 1112 if it is 25 minutes late. 1038 till 1112 means a wait of **34 minutes**.

SAMPLE QUESTIONS

1 Bill gets to work at 9.55 pm, and comes off the night shift at 6.04 am the following day. What length of time has he worked overnight?

2 My flight from Gatwick takes off at 1436 and is scheduled to last 4 hours 35 minutes. When should we land?

3 My recipe book says roast turkey needs 20 minutes per pound plus 20 minutes. How long will it take to roast a 13 lb turkey? When should I put the oven on, if it takes half an hour to warm up and I want the turkey roasted by 12.30 pm for Christmas lunch?

4 I have two and three quarter hours free on my video tape. How much will be left after I record the late film, starting at 11.35 pm and finishing at 1.05 am, the following morning?

5 The four laps of a relay race were run in 1 min 12 s, 1 min 3 s, 58 s and 1 min 5 s. How long did the race take altogether and what was the average time for each lap?

6 Using the Brightlington–Poolsberry timetable find: (a) the time of the last train from Brightlington which will reach Poolsberry before 8.30 pm, (b) the fastest journey time from Faringsham to Poolsberry in the morning, (c) the times of trains from Brightlington to Midlinghaven before 11.00 am and from Midlinghaven to Poolsberry after 12 noon and before 5.00 pm.

7 Today is Wednesday the 19th March 1986. What date is it: (a) next Wednesday, (b) two weeks tomorrow, (c) last Saturday? Give the dates of the second Sunday in March, April and May.

8 Mrs Bersil's washing machine has a programme which soaks for 25 minutes, washes for 12 minutes and finishes with 3 rinse cycles each taking 18 minutes. When will this programme finish if she starts it at: (a) 11.22 am, (b) 10.55 pm?

MEASURING
3.1 Units

The **Metric system** of units is used for many measurements nowadays, although there are instances when the **Imperial system** is still used. Much of the work in GCSE mathematics will use the metric system, but candidates must also be familiar with the Imperial units in common use and be able to convert between the two systems when required.

Metric units

Length
Length is based on the **metre**. The distance from the North Pole to the Equator was originally taken as 10 000 000 metres.

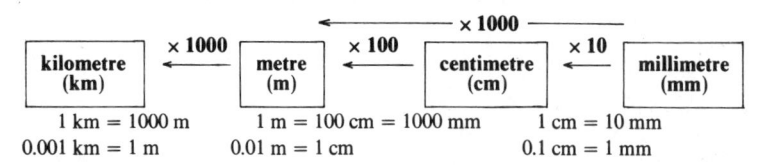

$$\text{kilometre (km)} \xleftarrow{\times 1000} \text{metre (m)} \xleftarrow{\times 100} \text{centimetre (cm)} \xleftarrow{\times 10} \text{millimetre (mm)}$$

$$\xleftarrow{\times 1000}$$

1 km = 1000 m   1 m = 100 cm = 1000 mm   1 cm = 10 mm
0.001 km = 1 m   0.01 m = 1 cm   0.1 cm = 1 mm

Mass
Mass is based on the **gram**, the mass of 1 cm^3 of water.

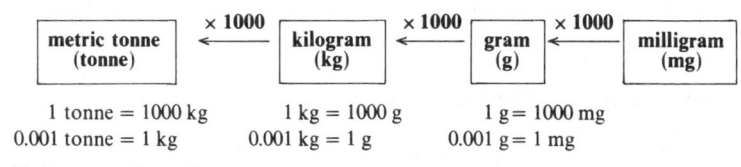

$$\text{metric tonne (tonne)} \xleftarrow{\times 1000} \text{kilogram (kg)} \xleftarrow{\times 1000} \text{gram (g)} \xleftarrow{\times 1000} \text{milligram (mg)}$$

1 tonne = 1000 kg   1 kg = 1000 g   1 g = 1000 mg
0.001 tonne = 1 kg   0.001 kg = 1 g   0.001 g = 1 mg

Volume or Capacity

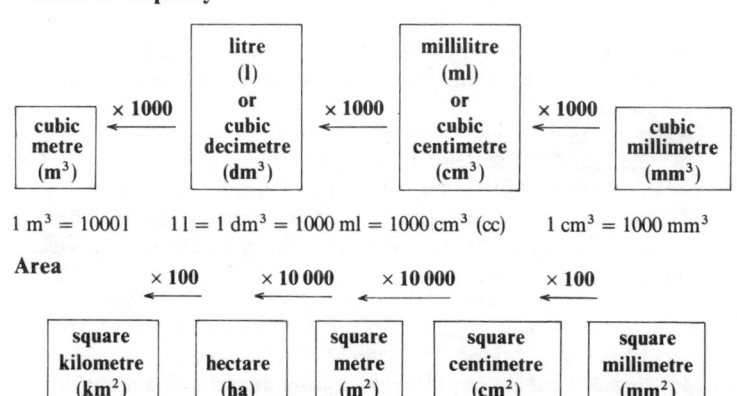

$$\text{cubic metre (m}^3\text{)} \xleftarrow{\times 1000} \text{litre (l) or cubic decimetre (dm}^3\text{)} \xleftarrow{\times 1000} \text{millilitre (ml) or cubic centimetre (cm}^3\text{)} \xleftarrow{\times 1000} \text{cubic millimetre (mm}^3\text{)}$$

1 m^3 = 1000 l   1 l = 1 dm^3 = 1000 ml = 1000 cm^3 (cc)   1 cm^3 = 1000 mm^3

Area

$$\text{square kilometre (km}^2\text{)} \xleftarrow{\times 100} \text{hectare (ha)} \xleftarrow{\times 10\,000} \text{square metre (m}^2\text{)} \xleftarrow{\times 10\,000} \text{square centimetre (cm}^2\text{)} \xleftarrow{\times 100} \text{square millimetre (mm}^2\text{)}$$

1 km^2 = 1000 m × 1000 m = 1 000 000 m^2 1 ha = 100 m × 100 m = 10000 m^2
1 m^2 = 100 cm × 100 cm = 10000 cm^2 1 cm^2 = 10 mm × 10 mm = 100 mm^2

Units 42

Time

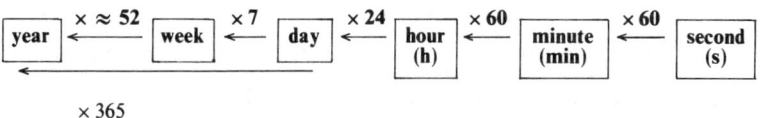

year ←×≈52— week ←×7— day ←×24— hour (h) ←×60— minute (min) ←×60— second (s)

× 365
or × 366 in leap year

Common imperial units
The numbers in the Imperial system are much harder to learn, but you should know this basic list of units and approximate Imperial-Metric **conversions**.

Length

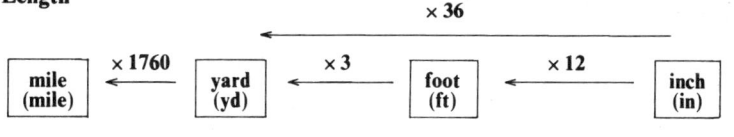

mile (mile) ←×1760— yard (yd) ←×3— foot (ft) ←×12— inch (in)

(×36 from yard to inch)

5 mile ≈ 8 km   1 yd ≈ 1 m   1 in ≈ 2.5 cm

Mass

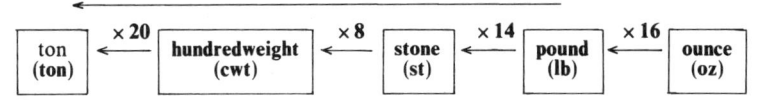

ton (ton) ←×20— hundredweight (cwt) ←×8— stone (st) ←×14— pound (lb) ←×16— ounce (oz)

(×2240 from ton to pound)

1 ton ≈ 1 tonne   1 lb ≈ 0.5 kg   1 oz ≈ 25 g

Capacity

gallon (gal) ←×8— pint (pt) ←×20— fluid ounce (fl oz)

1 gal ≈ 4.5 l   1 pt ≈ 0.5 l = 500 ml

Example   (*a*)   Convert 2 ft 4 in into metres, using 1 in ≈ 2.5 cm
(*b*)   London to Manchester 184 miles
Use 5 miles ≈ 8 km to convert this distance into kilometres.

(*a*)   2 ft 4 in = $(2 \times 12 + 4)$ in = 28 in ≈ (28×2.5) cm = 70 cm = **0.7 m**
(*b*)   5 mile ≈ 8 km, dividing by 5 gives   1 mile ≈ $(8 \div 5)$ km = 1.6 km
So 184 mile ≈ (184×1.6) km = 294.4 km ≈ **290 km (2SF)**

Example   Which is more, $\frac{1}{4}$ lb or 125 g of tea, given that 1 oz = 28.35 g?
How should 125 g be priced if $\frac{1}{4}$ lb of tea costs 42p?

$\frac{1}{4}$ lb = 4 oz = (4×28.35) g = 113.4 g, so **125 g** is the bigger amount.
$\frac{1}{4}$ lb of tea costs 42p so 113.4 g of tea would be priced at 42p.
Therefore   1 g of tea would be priced at 42p ÷ 113.4 = 0.370p,
and   125 g of tea should be priced at 0.370p × 125 = 46.25p ≈ **46p**.

43 Units

When a change of units is to be used 'in reverse' it is helpful to write out the conversion using scale factors, as shown in the next example.

Example Use the conversion $\boxed{1 \text{ pt} = 0.56 \text{ l}}$ to calculate how many pints are contained in a 3 litre wine box.

1 pt = 0.56 l means that pints may be converted into litres by **multiplying** by 0.56, but this also means that litres may be converted into pints by **dividing** by 0.56 as shown below.

$$\boxed{\text{pints}} \xrightleftharpoons[\div 0.56]{\times 0.56} \boxed{\text{litres}}$$

So 3 l = (3 ÷ 0.56) pt = **5.4 pt (2SF)**

Example An estate agent measures a room as 9 ft 9 in by 12 ft 10 in. Use the conversion $\boxed{1 \text{ m} = 39.4 \text{ in}}$ to write these lengths in metres. Carpet is sold from a roll 4 m wide. What length of carpet will be needed to completely cover this room?

The conversion 1 m = 39.4 in means $\boxed{\text{metre}} \xrightleftharpoons[\div 39.4]{\times 39.4} \boxed{\text{inch}}$

9 ft 9 in = (9 × 12 + 9) in = 117 in = (117 ÷ 39.4) m = 2.97 m (3SF)
12 ft 10 in = (12 × 12 + 10) in = 154 in = (154 ÷ 39.4) m = 3.91 m (3SF)
The dimensions of the room would be written **2.97 m by 3.91 m**.

The 4 m width of the roll will allow the full 3.91 m length of the room to be covered, so the length needed is 2.97 m, or **3 m** for cutting.

Currency conversion

When British currency is to be **exchanged** for foreign currency the current **conversion rate** is used. These rates, published daily in the newspapers, fluctuate depending on the strength of the particular currency involved.

France............11.65 francs	Germany............3.81 marks
Italy..............2555 lira	Holland............4.29 guilders
Spain..........221.75 pesetas	United States........1.39 dollars

The 'France' row means that:
1 pound sterling may be **exchanged** for **11.65 francs**.

Example Use the table above to convert:
(a) £300 into US dollars, (b) £58.65 into Deutschmarks.

(a) The conversion means $\boxed{\text{pounds sterling}} \xrightleftharpoons[\div 1.39]{\times 1.39} \boxed{\text{US dollars}}$

So £300 = (300 × 1.39) dollars = **417 dollars**.

(b) £58.65 = (58.65 × 3.81) marks = (223.4565) marks = **223.46 marks**, rounding to a whole number of pfennings, (100 pfennings = 1 mark)

When converting back into pounds sterling from a foreign currency, the conversion rate back into pounds will give a lower rate of return.

Example A family, travelling to Sweden, change £500 into kronor using the rate £1 = 11.44 kronor. How many kronor will they receive? On their return journey they change their remaining Swedish money, 1550 kronor, for £134.20 sterling. What exchange rate have they been given?

£500 = (500 × 11.44) kronor = **5720 kronor**.

On return they receive £134.20 for 1550 kronor, so dividing by 134.20,

£1 = (1550 ÷ 134.20) kronor = **11.55 kronor**.

SAMPLE QUESTIONS

1. Rewrite the first quantity in the second units: (a) 2.3 cm, m, (b) 56 kg, g, (c) 0.75 km, m, (d) 50 mg, g, (e) 225 ml, cm^3, (f) 34 l, m^3, (g) 5 yd 2 ft, ft, (h) 9 st 13 lb, lb, (i) 5 fl oz, pt.
2. Express the capacity of a glass flask, given as 255 cc, in litres.
3. Total these weights in kilograms: (a) 35 kg, 2.6 tonne, 0.54 kg, (b) 38 tonne, 450 kg, 23 tonne, (c) 355 g, 28.5 kg, 22 500 mg.
4. Calculate the area of a rectangular field, 350 m wide and 0.62 km long: (a) in square metres, (b) in square kilometres, (c) in hectares.
5. Convert the first quantity into the second units: (a) 5 lb, kilograms, (b) 3 in, centimetres, (c) 1500 yd, kilometres, (d) 200 gall, litres.
6. Use the conversion $\boxed{1\,l = 1.759\,pt}$ to write: (a) 30 litres in pints, (b) 24 pints in litres, (c) $\frac{1}{3}$ pint in millilitres.
7. The world's heaviest man weighed 485 kg when he died. Use $\boxed{1\,kg = 2.21\,lb}$ to write his weight in: (a) lb, (b) st and lb.
8. Use the table of currency conversions given opposite to change: (a) £35 into pesetas, (b) £2500 into guilders, (c) 500 francs into lira.
9. The first American moon landing has been estimated to have cost approximately 20 000 000 000 US dollars. How much is this in pounds sterling, if £1 = 1.395 dollars?
10. Use $\boxed{1\,lb = 0.45\,kg}$ to calculate the equivalent cost of a 1.5 kg bag of flour if a 3 lb bag costs 39p.
11. Celsius (°C) → $\boxed{\div 5}$ → $\boxed{\times 9}$ → $\boxed{+32}$ → Fahrenheit (°F) Use the flow diagram to convert, (a) 36.9°C to °F, (b) 72.5°F to °C.
12. Wine is often sold in 70 cl capacity bottles. (1 cl = 0.01 l). Calculate how many glasses may be filled from a case of 12 such bottles if each glass holds 2 fluid ounces. Use $\boxed{1\,\text{pint} = 0.568\,\text{litres}}$

MEASURING
3.2 Length

3

Estimation

When estimating distance, a **reference length** is needed. Common reference lengths are the height of a man or doorway often taken as 2 m, the height of a room 3 m, or the width of a hand 10 cm.

It is important not to overstate **the accuracy of an estimated length**, and usually an accuracy of 1 significant figure is sufficient.

Example Figure 1 shows an accident report map, marking roughly the path of a car which collided with a bus and then hit a tree. Assuming that the car is 14 ft long estimate the length of the bus and the distance travelled by the car between hitting the bus and stopping at the tree.

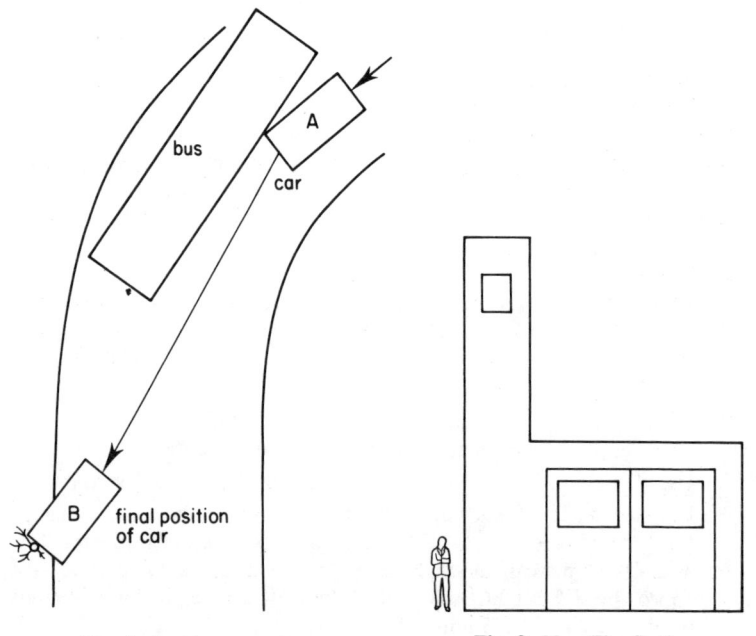

Fig. 1 Accident report.

Fig. 2 New Fire Station.

The length of the car on the diagram is measured as 13 mm, so an approximate scale would be 1 mm to a 1 foot.
The bus measures 4 cm in length so its length is estimated as **40 ft**.
The distance between the bus and the tree is measured as 6.2 cm so an estimate is 62 ft, which is **60 ft** to 1 significant figure.

Length 46

Example Figure 2 shows a scale diagram of a new Fire Station. Estimate the height of the tower and the length of the building.

The height of the fireman may be estimated as 2 m, to the nearest metre. Measuring from the diagram the fireman is 1 cm and the tower is 5 cm high, and so the height of the tower may be estimated at 5×2 m = **10 m**.
The length of the building measures 3.9 cm on the diagram so may be estimated as 3.9×2 m = 7.8 m = **8 m** to the nearest metre.

Reading linear scales

Example Use the scale to read off the values shown by the markers.

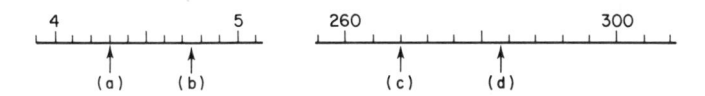

On the first scale there are 10 small divisions between 4 and 5, so each division is 0.1. So marker (a) is pointing to the value **4.3**.
Marker (b) is halfway between 4.7 and 4.8 so the value is **4.75**.
On the second scale there are 10 divisions between 260 and 300, so each division is 40/10 = 4. So marker (c) is pointing to the value **268**.
Marker (d) is nearer to 284 than 280, so the value is approximately **283**.

Scale diagrams

A **scale diagram** is a drawing in which all the lengths are drawn accurately to scale. The scale may be marked like a ruler, or stated in ratio form. For instance, the scale 1 cm to 40 m, means 1 centimetre on the diagram represents a true length of 40 metre.
The table shows equivalent scale lengths and true lengths for this scale.

Scale length (cm)	0	0.5	1	1.5	2	2.5	3	3.5	4	4.5
True length (m)	0	20	40	60	80	100	120	140	160	180

Example The Town plan shown in Figure 3 (see p. 47) is marked with a grid of squares of side 200m. Calculate the distance (a) from the station to the Red Lion, (b) from the garage to the hospital, (c) from the car park to the cinema via the town centre, (d) Find the area of the playing field.

The grid squares have edge 2 cm, so the scale is 2 cm : 200 m, 1 cm : 100 m.
(a) From the station to the Red Lion is 2 cm, representing **200 m**.
(b) From the garage to the hospital is 5 cm, representing **500 m**.
(c) From the car park to the town centre is 4 cm and from the town centre to the cinema is a further 3 cm, total 7 cm, representing **700 m**.
(d) The field measures 1.5 cm × 2.5 cm, representing 150 m × 250 m, an area of 37 500 m^2, **40 000 m^2 (1SF)**.

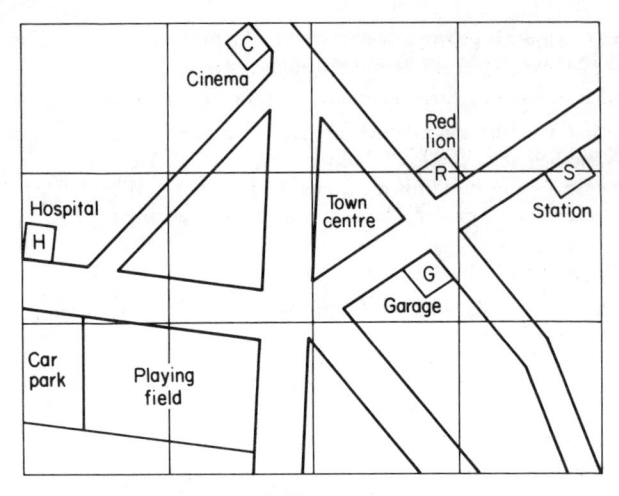

47 Length

Fig. 3 Town centre.

Example Figure 4 shows the existing house and garage and the proposed new garage at 1, Field Close. Calculate the dimensions of the new garage, and the length of the drive both before and after the new garage is built.

Fig. 4 New garage.

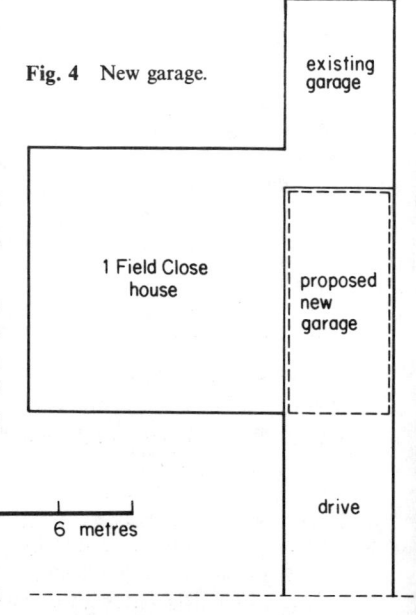

The divisions on the marked scale are 1 cm apart, each representing 2 m. The new garage measures 3 cm long by 1.5 cm wide on the diagram, which represents true dimensions of **6 m by 3 m**.

The drive measures 5.4 cm before and 2.4 cm after the new garage is built. 1 cm represents 2 m, so 0.1 cm represents 0.2 m, 0.4 cm represents 0.8 m. The length of the drive will be reduced from **10.8 m** to **4.8 m**.

The scale length chart is shown below.

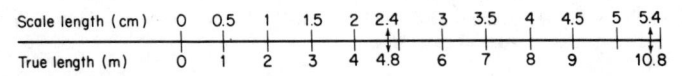

Scale length (cm)	0	0.5	1	1.5	2	2.4	3	3.5	4	4.5	5	5.4
True length (m)	0	1	2	3	4	4.8	6	7	8	9		10.8

Length 48

Drawing scale diagrams
Use the following hints when drawing a scale diagram.
If you have to *choose a scale* make sure that the diagram will *fit* on your paper and almost *fill* it. *Sketch a rough outline* of your diagram first on another piece of paper so that the final versinn is centred on the paper.
Write down the *scale* you are using clearly at the top of your diagram.
Calculate the corresponding scale lengths using a **conversion table**.

SAMPLE QUESTIONS
1 Estimate the width of the road in Figure 1.
2 Estimate the height and width of the pair of double doors of the Fire Station in Figure 2.
3 Calculate the dimensions of (a) the house, (b) the existing garage in Figure 4. No. 1, Field Close is a two storey house with the same floor area on both floors. Calculate: (c) the existing floor area (house and garage), (d) the new floor area (including the new garage), (e) the percentage increase in floor area.
4 Calculate the true distance between the following places drawn in Figure 5. (a) London, Paris, (b) Berne, Rome, (c) London, Rome.

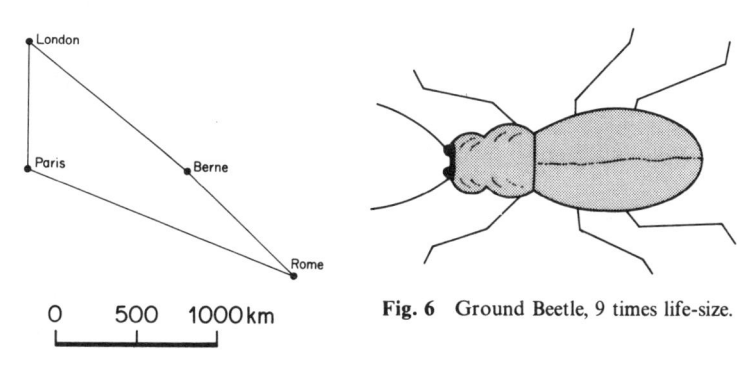

Fig. 5 Airline routes.

Fig. 6 Ground Beetle, 9 times life-size.

5 Using the scale diagram drawn in Figure 6 find (a) the length of the beetle's body, (b) the maximum width of the beetle's body, (c) the average length of the beetle's legs.
6 From a diagram with scale 1 cm to 25 cm, calculate the true length of the following scale length: (a) 3 cm, (b) 11 cm, (c) 0.2 cm, (d) 3.6 cm, (e) 4.1 cm, (f) 83 mm.
7 Calculate the scale length needed to represent 4.50 m on a drawing with the following scale: (a) 1 cm to 1 m, (b) 1 cm to 2 m, (c) 1 mm to 4 m, (d) 2 cm to 1 m, (e) 5 cm to 1 m, (f) 3 cm to 0.5 m.

MEASURING
3.3 Scale

Scale ratios

Example A design of a new kitchen is drawn to a **scale** of 1:20. The length of the kitchen is 18 cm and the new sink unit is 7.5 cm on the plan. What are the true dimensions? The new hob unit measures 50 cm by 60 cm. What size should it be drawn on the plan?

The **ratio of lengths** is 1:20, $\boxed{\text{scale length} \underset{\div 20}{\overset{\times 20}{\longleftrightarrow}} \text{true length}}$

so the true length of the kitchen will be 18 cm $\times$ 20 = 360 cm = **3.6 m**, and the true length of the sink unit will be 7.5 cm $\times$ 20 = 150 cm = **1.5 m**. The dimensions of the new hob unit must be divided by 20 to fit the scale, so they will be 50 cm $\div$ 20 by 60 cm $\div$ 20 = **2.5 cm by 3 cm**.

The Landranger Series of maps produced by the Ordnance Survey are drawn to a **scale** of 1:50 000. 1 cm on the map is 50 000 cm = 500 m = 0.5 km on the ground. The **length scale factor** is 50 000, which means that any length on the ground is 50 000 times longer than the corresponding length on the map.

Example Oxford and Witney are 28 cm apart on a Landranger O.S. map. How far apart are they on the ground? If Bicester is 18 km from Oxford, how far apart will these towns be drawn on the same map?

$\boxed{\text{map distance} \underset{\div 50\,000}{\overset{\times 50\,000}{\longleftrightarrow}} \text{true distance}}$

Oxford to Witney is 28 cm $\times$ 50 000 = 1 400 000 cm = 14 000 m = **14 km**.
Oxford to Bicester is drawn as
18 km $\div$ 50 000 = 0.00036 km = 0.36 m = **36 cm**.

Scale changes on area and volume
*Two objects are **similar** if one is an exact enlargement of the other.
When a scale of 1:n is being used for an enlargement:

the length scale factor is n
the area scale factor is
  $n \times n = n^2$
the volume scale factor is
  $n \times n \times n = n^3$

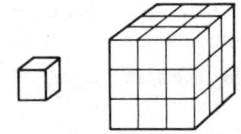

Fig. 1 Length $\times 3$ area $\times 9$ volume $\times 27$.

Example Verify that the area scale factor for the new hob unit in the first example is 400.

With a **scale** of 1:20 the **area scale factor** should be 20 $\times$ 20 = 400.
On the plan the unit is drawn as 2.5 cm by 3 cm, so has an area of 7.5 cm^2.
The real hob unit has dimensions 50 cm by 60 cm, with an area of 3000 cm^2.
So the area scale factor is 3000 $\div$ 7.5 = **400**.

Scale 50

With a scale of 1:20 it is easy to think that all measurements have the same scale, but **area**, **volume**, and **mass** will have different scale factors.

* **Example** An exact scaled down replica of a stone statue has height 19.5 cm and mass 2.3 kg. If the full size statue is 1.56 m high find the scale used and the mass of the statue.

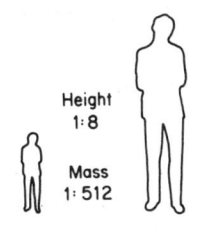

Working in centimetres, the scale factor is $156 \div 19.5 = 8$, so the scale is **1:8**. The **volume scale factor** will be $8 \times 8 \times 8 = 8^3 = 512$. The masses of the statues will be proportional to their volumes and so the mass of the full size statue will be
512×2.3 kg $= 1177.6$ kg $=$ **1180 kg (3SF)** (Figure 2).

Height 1:8
Mass 1:512

Fig. 2 Similar statues.

SAMPLE QUESTIONS

1 A model of a ship is 20 cm long, 5 cm wide and has a mast 12.5 cm high. If the real ship is 70 m long, how wide is it and how high is its mast?
2 Copy and complete the table

ground distance	10 km	15 km		1 km		600 m
map distance	8 cm		15 cm		2 mm	

3 On a map with a scale of 1:25 000, what lengths on the ground do the following represent: (*a*) 3 cm   (*b*) 14 cm   (*c*) 4 mm   (*d*) 0.8 m.
4 Draw a rectangle 30 mm by 5 mm and a scale factor 3 enlargement of it. What is the perimeter of the larger rectangle? Give the area scale factor.
5 How many square centimetres are there in 1 m^2? How many cubic millimetres are there in 1 km^3?
*6 A photograph of a tower is enlarged so that the new image of the tower is 4 times as tall. If the cost of photographic paper for the enlargement is £2.56 how much did the paper for the original photograph cost?
*7 Two similar cans of soup have heights 8 cm and 12 cm. The smaller can holds 250 ml. How much does the larger can hold?
8 Using a scale of 1:50 draw a net of a cuboid of length 8 m, width 5 m and height 7.5 m. If the net was made up into a cuboid how many would fit into the large cuboid?
9 Figure 3 shows a plan of a house drawn to a scale 1:150.
(*a*) What are the true dimensions of the kitchen, lounge and hall?
(*b*) What is the ground floor area?

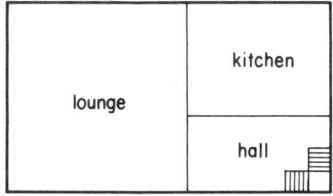

Fig. 3 House plan, scale 1:150.

MEASURING
3.4 Angle

Angles
The basic unit of turning, the **angle**, is usually measured in **degrees**. Figure 1 shows the common angles:

(a) **full turn** = 360 degrees, written 360°,
(b) **half turn** = 180°, the angle on a line,
(c) **quarter turn** = 90°, a **right angle**,
(d) **eighth turn** = 45°, **half a right angle**,
(e) **sixth turn** = 60°, angle in an **equilateral** triangle,
(f) **twelfth turn** = 30°, half of 60° or one third of a right angle.

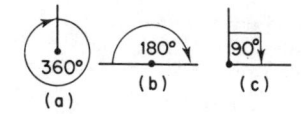

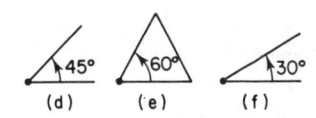

Fig. 1 Common angles.

Angle work may also involve clock faces, compass directions or bearings. **Bearings** are measured **clockwise from North** and written as a 3 digit number. The direction East would be on a bearing of 090°.

Example Find the angle turned in the following cases: (a) start facing North, turn clockwise until facing South West; (b) by the large hand, (c) by the small hand of a clock, between 9.00 pm and 9.20 pm; (d) start on a bearing of 030° turn anticlockwise to a bearing of 340°.

Draw a simple diagram in each case (see Figure 2), marking the angle needed by an arc.

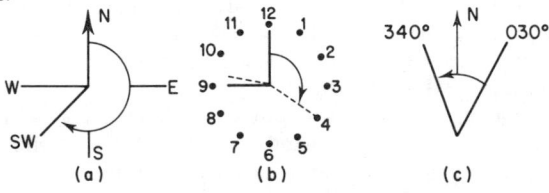

Fig. 2 Calculating angles.

(a) Between North and South is a half turn, 180°, and between South and South West is half of a quarter turn, which is $\frac{1}{2}$ of 90° = 45°. The total angle turned is 180° + 45° = **225°**.

(b) Turning from the 12 to the 3 is a quarter turn, 90°, and between the 3 and the 4 is one third of a quarter turn, which is $\frac{1}{3}$ of 90° = 30°. The total angle turned is 90° + 30° = **120°**.

(c) Between 9.00 pm and 10.00 pm the small hand will turn one twelfth of a full turn, which is $\frac{1}{12}$ of 360° = 30°. Between 9.00 pm and 9.20 pm is 20 minutes, one third of an hour, so the small hand will only turn one third of this angle, $\frac{1}{3}$ of 30° = **10°**.

Angle 52

(d) Turning anticlockwise between 030° and 000° (due North), is 30°, and between North and 340° is a further 20°, a total of 30° + 20° = **50°**.

Angle sums and parallel line properties

The angles at a point add up to 360°

In Figure 3, $x° + 290° = 360°$
and so $x° = 70°$

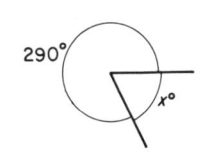

Fig. 3 Angle at a point.

The angles on a line add up to 180°

In Figure 4, $y° + 30° + 110° = 180°$
and so $y° = 40°$

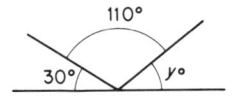

Fig. 4 Angle on a line.

When two lines cross, two pairs of equal **vertically opposite** angles are made.
In Figure 5, the angle $a°$ is opposite 40° so $a° = 40°$, and $b° = c° = 140°$

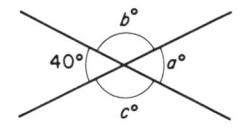

Fig. 5 Opposite angles.

A **transverse** line cutting two parallel lines produces equal **corresponding** and **alternate** angles.
In Figure 6, $d°$ is a corresponding angle to 72°, so $d° = 72°$, and $e°$ is an alternate angle to 72°, so $e° = 72°$. All the other angles at these points are 72° or 108°.

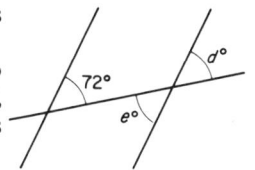

Fig. 6 Angles on parallel lines.

Estimating angles

Angles may be estimated by comparing with the standard angles, 360°, 270°, 180°, 90°, or for smaller angles the fractions of a right angle, 45°, 30° and 60°.

Example Estimate the angles marked in Figure 7: (a) between the crane jib and the horizontal, (b) turned by the window, (c) turned by the swing door, (d) the bearing of the mast from the church.

(a) The angle is less than a right angle, 90°, but more than an eighth turn 45°. For greater accuracy, compare with the angle of an equilateral triangle, 60°, and the crane angle is just a little more, say **70°**.
(b) The window has turned through almost a half turn 180°, so consider the angle needed to complete the half turn. This angle is approximately half

53 *Angle*

of 45°, say 20°. So the window has turned through $180° - 20° = \mathbf{160°}$.

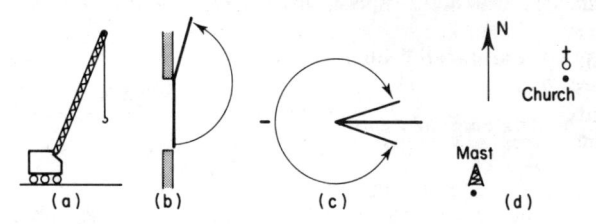

Fig. 7 Estimate the angles marked.

(c) The swing door has turned through more than a three quarter turn, 270°, but less than a full turn 360°, so consider the angle needed to complete the full turn. This angle is slightly less than 45°, say 40°, so the swing door has turned through $360° - 40° = \mathbf{320°}$.

Fig. 8

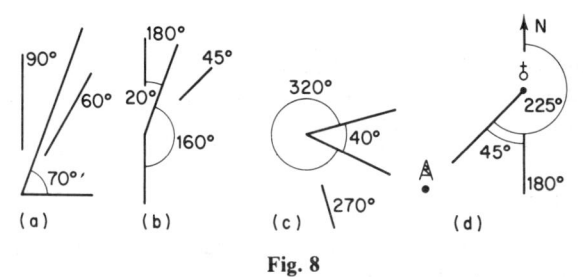

Figure 8 shows the required angle and some of the reference angles used for comparison.

(d) Draw a North line from the church and mark the angle required to turn clockwise from North to face the mast. This angle is more than a half turn, 180°, so mark the 180° angle and consider the remaining angle. This angle is approximately 45°, so the bearing of the mast from the church is $180° + 45° = \mathbf{225°}$.

Measuring angles

Angles are measured using a **protractor**, either semicircular marked 0° to 180°, or circular marked 0° to 360°. A circular protractor has advantages for angles greater than 180°, in bearings work for instance.

Place the protractor so that the centre and the 0° mark lie on one line and then read off the angle by following the scale round to the other line. There are frequently two scales marked clockwise and anticlockwise and it is a common error to read the wrong scale.

Make an estimation of the angle first and then compare this with the protractor reading to avoid such errors.

Angle 54

SAMPLE QUESTIONS
1. Calculate the angle turned in degrees: (a) two thirds of a full turn, (b) one tenth of a full turn, (c) a sixth of a right angle, (d) between West and North East, clockwise, (e) between bearings 170° and 275°, anticlockwise, (f) by the large hand between 2.15 am and 3.55 am, (g) by the small hand between 12 noon and 9.30 pm.
2. Find the unknown angle or angles marked in Figure 9:

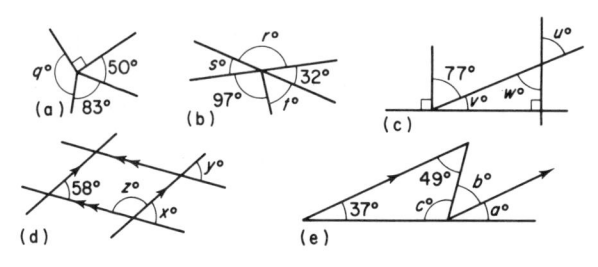

Fig. 9

3. Estimate: (a) the angle which the greenhouse roof makes with the horizontal, (b) the angle which the guy ropes of the tent make with the vertical, (c) the angle of depression (below horizontal) of the telescope on the clifftop, (d) the angle between the roof lines (see Figure 10).

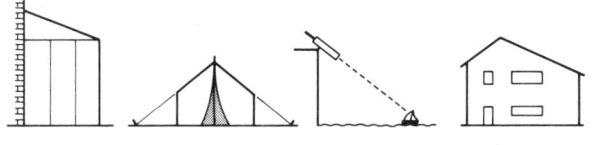

Fig. 10 (a) greenhouse (b) tent (c) telescope (d) roof lines.

4. A yacht sailed the course shown in Figure 11. Copy and complete the table showing distances and bearings on the first 3 legs of the trip. From the diagram work out the final distance and bearing needed to return to the start point.

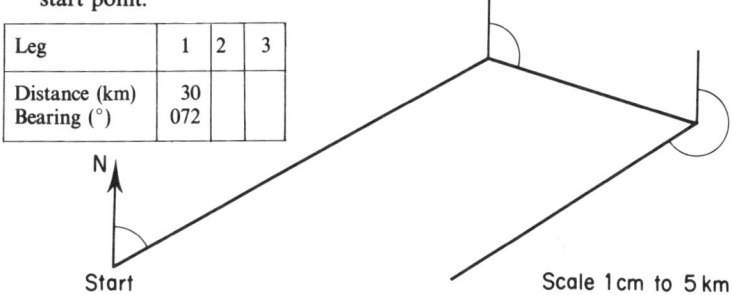

Leg	1	2	3
Distance (km)	30		
Bearing (°)	072		

Fig. 11 Sailing a course.

MEASURING
3.5 Perimeter   3

Perimeter of rectilinear shapes
The total distance around a flat shape is called the **perimeter**.

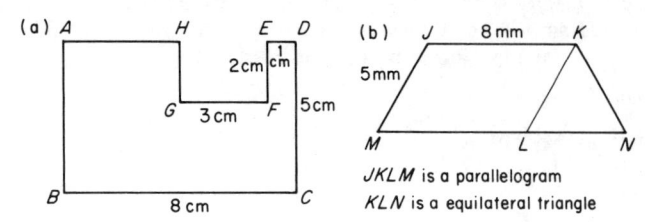

Fig. 1 Perimeter of simple shapes.

JKLM is a parallelogram
KLN is a equilateral triangle

Example Calculate the perimeter of the shapes drawn in Figure 1.

(a) When finding the perimeter of a shape built up from rectangles begin by marking every part of the perimeter with its length. So using CD, $AB = 5$ cm, $AH = (8-3-1)$ cm $= 4$ cm and $GH = 2$ cm, like EF. Then the perimeter is $(5+8+5+1+2+3+2+4)$ cm $= $ **30 cm**.

(b) Since $JKLM$ is a parallelogram, $ML = JK = 8$ mm, $KL = JM = 5$ mm, and since KLN is an equilateral triangle, $LN = KN = KL = 5$ mm. So $MN = 13$ mm. Then the perimeter of $JKNM$ is $(8+5+13+5)$ mm $= $ **31 mm**.

Pythagoras' theorem
This result is used to calculate the third side of a right-angled triangle from the other two sides. In Figure 2
$$a^2 + b^2 = c^2 \quad \text{or}$$
$$BC^2 + AC^2 = AB^2$$

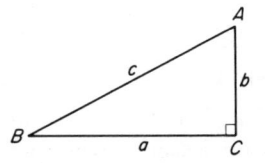

Fig. 2 Pythagoras' theorem.

Example Find the longest side of a right-angled triangle with shorter sides 3.2 cm and 4.5 cm.

Use the notation of Figure 2, with $BC = 3.2$ cm and $AC = 4.5$ cm.
Then $\qquad\qquad BC^2 + AC^2 = AB^2$
So $\qquad\qquad (3.2)^2 + (4.5)^2 = AB^2$
Using a calculator $\qquad\qquad 30.49 = AB^2$
Square root both sides $\qquad\qquad 5.522 = AB$
So the longest side is **5.5 cm (2SF)**.

*Example Figure 3 shows a fishing rod *RS* of length 2.3 m, with 5.8 m of line cast out to a fish at *F*. The tip of the rod *S* is 1.3 m above the water *W*. Calculate the distance of the fish from the base of the rod.

The diagram shows two right-angled triangles, RSW and FSW.

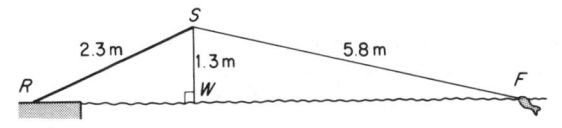

Fig. 3 Fishy problem.

In triangle RSW

$RW^2 + WS^2 = RS^2$
$RW^2 + (1.3)^2 = (2.3)^2$
$RW^2 + 1.69 = 5.29$
$RW^2 = 3.6$
$RW = 1.90$ (3SF)

In triangle FSW

$FW^2 + WS^2 = FS^2$
$FW^2 + (1.3)^2 = (5.8)^2$
$FW^2 + 1.69 = 33.64$
$FW^2 = 31.95$
$FW = 5.65$ (3SF)

Then the distance from the rod to the fish is $(1.90 + 5.65)$ m = **7.6 m (2SF)**.

SAMPLE QUESTIONS
1 Calculate the perimeter of a football pitch, 54 m by 23 m.
2 Calculate the perimeter of the shapes in Figure 4:

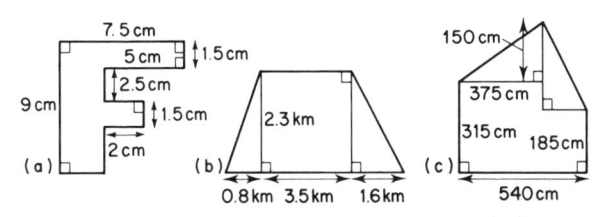

Fig. 4 Perimeter of shapes.

3 Find the length of a ladder which reaches a point 4.7 m up a vertical wall when resting with its feet level 1.8 m from the wall.
4 A children's play area has been made by marking out a corner of a rectangular field as shown in Figure 5. Calculate the distance *FG*, and the cost of erecting a perimeter fence at £4.50 per metre round the area.

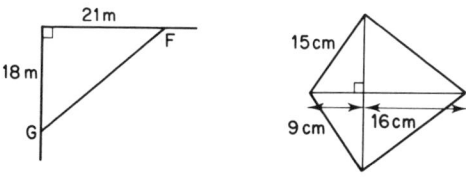

Fig. 5 Play area.   **Fig. 6** Kite.

*5 Calculate the perimeter of the kite drawn in Figure 6.

MEASURING
3.6 Circles   3

A **circle** is the set of points in a plane which lie at a constant distance, the **radius**, from a fixed point, the **centre**, see Figure 1.
The distance across a circle, called the **diameter**, is twice the radius, and around the edge, the perimeter, is called the **circumference**.

Circle formulae
For a circle, radius r, diameter d, circumference C, and area A,
$C = 2\pi r = \pi d$, $A = \pi r^2$,
where $\pi = 3$ (1SF)—for rough work,
$\pi = 3.14$ (3SF)—for most purposes,
$\pi = 3.1415927$—calculator π button.

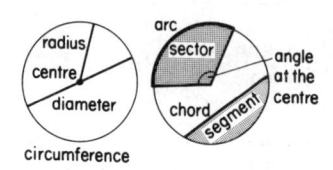

Fig. 1 Vocabulary of circles.

Example Calculate (a) the area of a coffee table 52.4 cm in diameter, (b) the circumference of a wheel of radius 23 cm.

(a) Use $A = \pi r^2$, with $\pi = 3.14$, and $r = \frac{1}{2}$ of 52.4 cm = 26.2 cm. The area is $3.14 \times (26.2)^2$ cm^2 = 2155.4216 cm^2 = **2160 cm^2** (3SF). Using the π button on a calculator gives an area of 2156.5149 cm^2, which rounds to the same final value. Press $\boxed{\pi}$ $\boxed{\times}$ 26.2 $\boxed{x^2}$ $\boxed{=}$

(b) Use $C = 2\pi r$, and the same values for π and r. The circumference is $2 \times 3.14 \times 26.2$ cm = 164.536 cm = **165 cm** (3SF). Using a calculator π button gives 164.61946 cm = 165 cm (3SF), as before.

Example A gardener is considering laying a lawn on a rough piece of ground 5 m across. Calculate the area of turf and the length of metal edging strip required if he lays (a) a circular lawn, (b) a square lawn 5 m across. Turf costs 48p per square metre and edging costs 78p per metre. Calculate the total cost in each case.

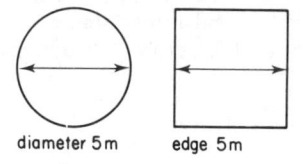

Fig. 2 (a) Circular or (b) square lawn. 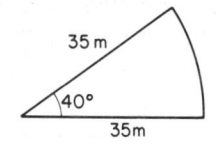   **Fig. 3** Safety zone.

Begin by drawing a diagram to show the two cases, see Figure 2.

(a) The radius of the circular lawn will be $\frac{1}{2}$ of the diameter, $\frac{1}{2}$ of 5 m = 2.5 m, and take $\pi = 3$, remembering this is a low estimate for π. The area will be $3 \times (2.5)^2$ m^2 = 18.75 m^2, and so

the gardener should order **20 m²** of turf. The circumference will be $2 \times 3 \times 2.5$ m = 15 m, so order **16 m** of metal edging strip.
(b) The area of the square lawn will be (5×5) m² = **25 m²**, and the circumference will be 4×5 m = **20 m**.

Costs Turf Edging Total	(a) 20 m² at 48p 16 m at 78p	£9.60 £12.48 £**22.08**	(b) 25 m² at 48p 20 m at 78p	£12.00 £15.60 £**27.60**

Arcs and sectors

*The **length of an arc** and the **area of a sector** may be calculated as a fraction of the **circumference** and **area** of the circle.

*Example A shot-put safety zone is in the shape of a sector of a circle, radius 35 m, with an angle of 40° at the centre. Calculate, to 2 SF, (a) the area of the safety zone, and (b) the length of rope needed to erect a simple rope barrier around the edge of the zone.

Figure 3 shows the zone, marking the radius and the angle at the centre.
(a) The area of the sector, angle 40°, of a circle is $\frac{40}{360}$ of the area of the complete circle. $\frac{40}{360} = \frac{1}{9}$, and so the zone has area $\frac{1}{9}$ of (3.14×35^2) m² = $\frac{1}{9}$ of 3846.5 m² = **430 m²** (2SF).
(b) The length of the arc of a sector, angle 40°, of a circle is $\frac{40}{360}$ of the circumference of the complete circle. So the length of the curved portion of the edge of the zone is $\frac{1}{9}$ of $(2 \times 3.14 \times 35)$ m = $\frac{1}{9}$ of 219.8 m = **24.4 m** (3SF). The total length of the edge of the safety zone will be $24.4 + (2 \times 35)$ m = 94.4 m = 94 m (2SF), needing **94 m** of rope.

SAMPLE QUESTIONS
1 Calculate, accurate to 3 SF, the circumference and area of the circle with the following dimensions; (a) radius 165 mm, (b) radius 8.7 m, (c) diameter 25 km, (d) diameter 0.025 cm, (e) diameter 3 inches.
2 Calculate the length of neon tubing needed to make an Olympic symbol with 5 circles of 4.2 m diameter.
*3 Calculate the length of curved curtain track needed for a bay window in the shape of a sector, angle 160°, of a circle radius 2.8 m.
4 A running track 3 m wide has two straights of length 104.5 m and two semicircular curved ends of inner radius 30.4 m. Calculate the overall length of the inner edge and the area enclosed by the track. Calculate the difference between the lengths of the inner and outer edge of the track.
*5 A baseball ground is shaped like a sector, angle 95°, of a circle of radius 105 m. Calculate its: (a) area, (b) perimeter fence length.
*6 A model train track has: 18 straight sections each of length 420 mm; 6 curves in the shape of an arc, angle 30°, of a circle radius 540 mm; 9 curves in the shape of an arc, angle 20°, of a circle radius 720 mm. Calculate the greatest possible length of a train circuit.

MEASURING
3.7 Area

3

The **area** of a shape is the amount of space within its boundary. Area is measured in square units, for instance **square metres**, written **m^2**.

Estimation of area
Area may be estimated by **counting squares** as shown in the next example.

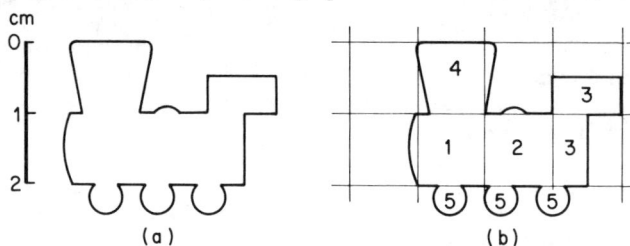

Fig. 1 Counting squares.

Example Figure 1(a) shows a simple train motif to be sewn on to a child's jumper. Use a 1 centimetre grid to calculate the area of the design.

Each square has area 1 cm^2. The squares and part squares within the boundary have been numbered in Figure 1(b) in order to count them. Two half squares have been labelled 3 and the three wheels have been numbered 5 since the area of each is roughly one third of a square. The approximate area is therefore 5 square centimetres, written **5 cm^2**.

Area formulae

$$\text{Area of a rectangle} = \text{length} \times \text{breadth}$$
$$\text{Area of a square} = (\text{length})^2$$
$$\text{Area of a right-angled triangle} = \tfrac{1}{2}(\text{length} \times \text{breadth})$$
$$\text{Area of a general triangle} = \tfrac{1}{2}(\text{base} \times \text{height})$$

Figure 2 gives a simple example of each shape with the area grid marked.

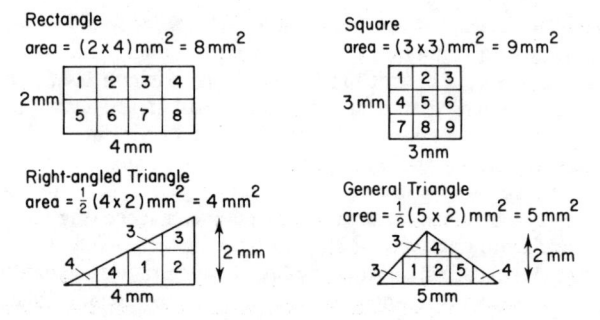

Rectangle
area = (2×4) mm^2 = 8 mm^2

Square
area = (3×3) mm^2 = 9 mm^2

Right-angled Triangle
area = $\tfrac{1}{2}(4 \times 2)$ mm^2 = 4 mm^2

General Triangle
area = $\tfrac{1}{2}(5 \times 2)$ mm^2 = 5 mm^2

Fig. 2

Area 60

The area of composite shapes may be calculated by splitting into standard shapes and using the formula for each one.

Example Find the area of (*a*) the lounge dining room drawn in Figure 3(*a*) and (*b*) the plot of land in Figure 3(*b*), accurate to 2 significant figures.

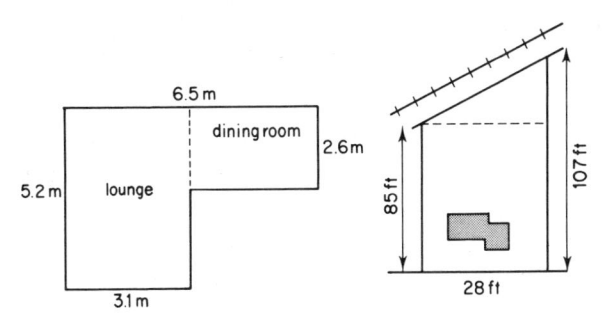

Fig. 3 (a) Lounge dining room. (b) Plot of building land.

(*a*) The lounge dining room is made up of two rectangles, with dimensions 5.2 m × 3.1 m and 3.4 m × 2.6 m, since the length of the dining room is (6.5 − 3.1) m, which is 3.4 m.
The area is $(5.2 \times 3.1) + (3.4 \times 2.6)$ m^2 = 24.96 m^2 = **25 m^2 (2SF)**.

(*b*) The trapezium shape may be split into a rectangle 28 ft × 85 ft, and a right-angled triangle, base 28 ft and height 22 ft.
The total area is $(28 \times 85) + \frac{1}{2}(28 \times 22)$ ft^2 = 2688 ft^2 = **2700 ft^2**.

Example Find the area of the brickwork facing of a railway tunnel which has the shape of a rectangle, 12.5 m wide and 5.3 m high, with a semicircle, radius 3.8 m, removed for the tunnel entrance. Calculate the approximate number of facing bricks used in its construction if each brick's face measures 25 cm by 8 cm. $\boxed{\text{Area of a circle} = \pi r^2, \pi = 3.14}$.

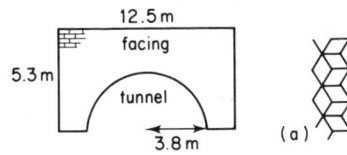

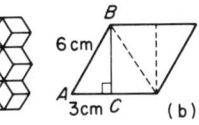

Fig. 4 Tunnel facing. **Fig. 5** Patchwork design.

Figure 4 shows a simple sketch of the brickwork.
The area of the rectangle is (12.5×5.3) m^2 = 66.25 m^2.
The area of the semicircle to be removed is $\frac{1}{2}$ of (3.14×3.8^2) m^2, which is $\frac{1}{2}$ of 45.34 m^2 (4SF) = 22.67 m^2 (4SF).
Then the area of the brickwork facing is $(66.25 - 22.67)$ m^2 = **43.6 m^2 (3SF)**.
Working in metres to make the calculations easier, each brick has face area (0.25×0.08) m^2 = 0.02 m^2.

61 *Area*

To find the number of bricks used we calculate how many 0.02's make 43.6, which is $43.6 \div 0.02 = 2180$, and so **2200 bricks (2SF)**, are required.

*Example A patchwork quilt, 2.1 m by 1.5 m, is to be made using a 'tumbling block' pattern, a tessellation of 60°, 120° angle rhombuses, see Figure 5(a). Find the area of a 6 cm edge rhombus. How many will be needed to finish the quilt? If a 3 cm edge rhombus were used instead how many would be needed to finish the quilt?

Divide the 6 cm rhombus into two equilateral triangles of edge 6 cm, and then again into four right-angled triangles, see Figure 5(b). The height BC of the right-angled triangle ABC may be found by Pythagoras' theorem, see Perimeter, page 55, since $AB = 6$ cm and $AC = 3$ cm.
$BC^2 + AC^2 = AB^2$
 $BC^2 + 3^2 = 6^2$
  $BC^2 = 27$ and so $BC = 5.196$ cm
Then the area of triangle ABC is $\frac{1}{2}(3 \times 5.196)$ cm^2 = 7.794 cm^2 (4SF), and the rhombus has area (7.794×4) cm^2 = **31.2 cm^2** (3SF).
In centimetres, the area of the quilt is (210×150) cm^2 = 31 500 cm^2. So the number of patches needed is $31\,500 \div 31.2 =$ **1010 (3SF)**.

The edge of a 6 cm rhombus are twice the size of the edges of a 3 cm rhombus patch, so its area is 4 times as large. Therefore 4 times as many 3 cm patches are needed, **4040 patches**.

SAMPLE QUESTIONS
1 Figure 6 is a scale diagram of an industrial estate, with access road and 9 plots. Calculate the area of; (*a*) plot 6, (*b*) plot 1, (*c*) plot 2, (*d*) the access road, (*e*) the entire estate. What percentage of the area of the estate has been used for the access road?

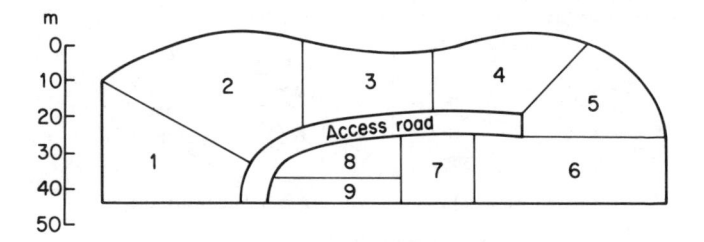

Fig. 6 Industrial estate.

2 Calculate the area of the shapes drawn in Figure 7; (a) the side wall of the house, (b) the glass on the side wall of the lean to greenhouse (excluding door), (c) the cross-section of the water pipe, $A = \pi r^2$, $\pi = 3.14$, (d) the kite.

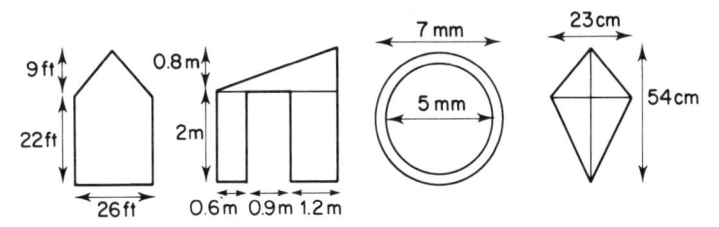

Fig. 7 (a) house wall   (b) greenhouse glass   (c) water pipe   (d) kite.

3 Calculate the cross-sectional area of a radiator key which is in the shape of a circle of diameter 9.00 mm with a square hole of side 5.00 mm, see Figure 8. Use $\boxed{A = \pi r^2, \pi = 3.14}$.

*4 Figure 9 shows a tiling pattern of octagons and squares to be used to cover a kitchen floor 4.2 m by 3.4 m. Calculate; (a) the area of the square tile, (b) the area of the octagonal tile (hint: split into 4 right-angled triangles, 4 rectangles and a central square). Find the total area of one of each shape tile and hence the approximate number of tiles of each shape needed to cover the kitchen floor.

*5 Figure 10 shows a simplified map of four of the states of America, giving the length in miles of some of the borders and the population, in millions, of each state in 1970. Calculate the area and population density (people per square mile) for each state at that time.

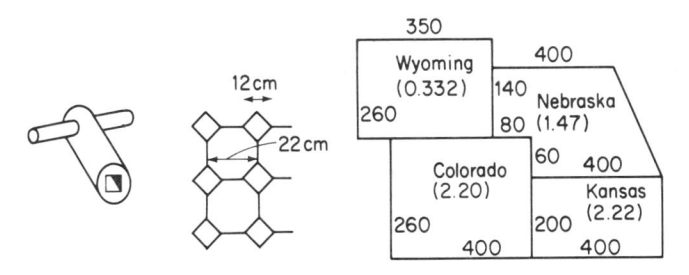

Fig. 8 radiator key.   **Fig. 9** Tiling pattern.   **Fig. 10** American states.

MEASURING
3.8 Nets and Surface Area   3

The **net** of a solid figure is a plane shape which can be folded to make the surface of the solid figure. The net may be used to **construct** the solid, to calculate its **surface area**, or to consider the **symmetry** of the solid. Figure 1 shows two different nets which both fold to make a cube, and one net each for a cuboid, a triangular based prism and a triangular based pyramid.

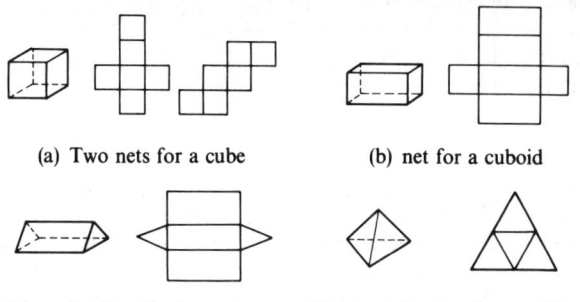

(a) Two nets for a cube (b) net for a cuboid

(c) triangular based prism net   (d) triangular based pyramid net

Fig. 1   Nets of simple solids

Example Calculate the length of the edge BE of the pyramid drawn in Figure 2(a). Draw a scale diagram of a net for the square based pyramid. Calculate: (a) the area of the face ABE, (b) the length of the edge CE, (c) the surface area of the pyramid. State the number of planes of symmetry of the pyramid.

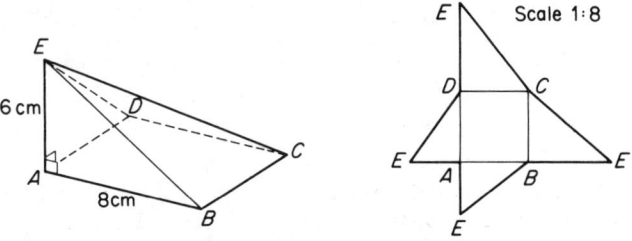

Fig. 2 (a) Square based pyramid   (b) Net for the pyramid

The triangle ABE is right-angled at A so use Pythagoras' result.

$AB^2 + AE^2 = BE^2,$
$8^2 + 6^2 = BE^2,$
so $100 = BE^2$, and the length of BE is **10 cm**.

Start the drawing of the net with the base $ABCD$, a square of edge 8 cm, scaled down to 1 cm, see Fig. 2(b). Then triangles ABE and ADE are right-angled at A with AE of length 6 cm, scaled down to 0.75 cm. Finally triangles BCE and DCE are right-angled at B and D respectively with BE and DE of length 10 cm, scaled down to 1.25 cm.

Nets and Surface Area 64

(a) The area of triangle ABE is $\frac{1}{2}(6 \times 8)$ cm^2 = **24 cm^2**.
(b) The length of CE may be found by using Pythagoras' again in triangle BCE. $\quad BC^2 + BE^2 = CE^2$,
$$8^2 + 10^2 = CE^2,$$
so $164 = CE^2$, and the length of CE is **12.8 cm** (3SF).
(c) The surface area of the pyramid is made up from square $ABCD$, two congruent triangles ABE and ADE, and two more congruent triangles BCE and DCE, with area $\frac{1}{2}(8 \times 10)$ cm^2 = 40 cm^2.
Total area is $(8 \times 8) + (2 \times 24) + (2 \times 40)$ cm^2 = **192 cm^2**.
The net has a line of symmetry AC, and the pyramid has just **one** plane of symmetry ACE.

Cylinders and cones

The net of solids with a curved surface cannot always be drawn, for instance, there is no possible net for a sphere. However the cylinder and cone both have a net, which allows the surface area of these figures to be calculated.

*Example Draw the net of a cylinder of radius r, height h, and find a formula for its surface area. Calculate the area of metal used to make a cylindrical can of height 8.2 cm and diameter 4.6 cm, assuming that no extra metal is needed to make the joins.

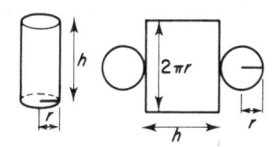

Fig. 3 (a) Net for a cylinder (b) Net for a cone

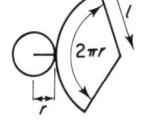

Figure 3(a) shows a net for the cylinder, based upon the standard net for any prism, with the base and top being circles of radius r, and the curved surface flattened into a rectangle. The height h of the cylinder gives the length of this rectangle and the width is the same length as the circumference of the base of the cylinder, $2\pi r$.

A formula for the **surface area of the cylinder** is $2 \times (\pi r^2) + 2\pi r \times h$, which may be simplified to $2\pi r^2 + 2\pi r h$.
Use the formula to calculate the area of metal needed for the can, with $h = 8.2$ cm, $r = \frac{1}{2}$ of 4.6 cm = 2.3 cm, and $\pi = 3.14$.
Area of can = $2 \times 3.14 \times (2.3)^2 + 2 \times 3.14 \times 2.3 \times 8.2$ cm^2
$\qquad\qquad = 151.662$ cm^2 = **152 cm^2** (3SF).

Figure 3(b) shows the net for a cone with base radius r and slant height l, and **the surface area** is $\pi r^2 + \pi r l$.

65 Nets and Surface Area

Plan and elevation

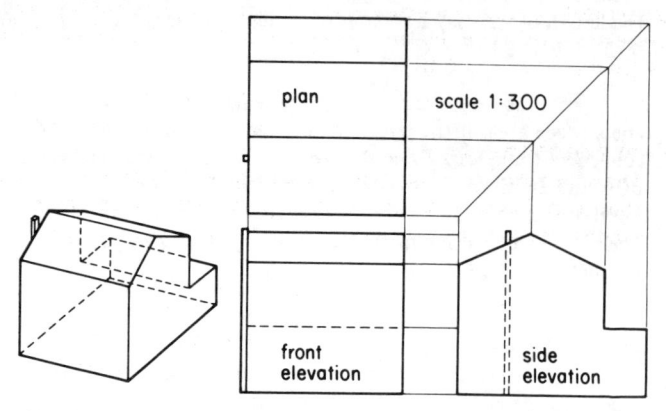

Fig. 4 (a) Perspective picture   (b) Plan and Elevations of house

An alternative way of drawing a solid figure is to draw a view from each of **three perpendicular directions**. Figure 4(b) shows the **plan, front elevation** and **side elevation** of the house drawn in 4(a). Solid lines are used for visible edges and dotted lies for hidden detail.

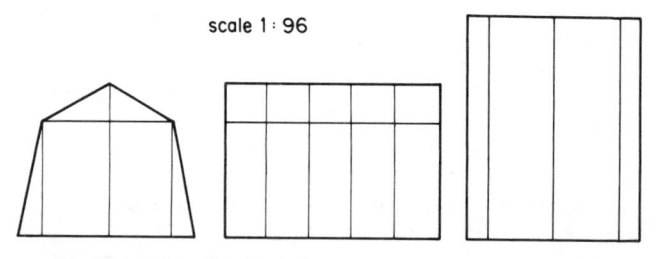

scale 1 : 96

Fig. 5 (a) Front and side elevation   (b) Plan of greenhouse

Example Figure 5(a) shows the front and side elevation of a greenhouse. Draw the plan and take measurements from the three diagrams in order to find approximately the area of glass needed to glaze the complete greenhouse.

The front and side elevations show that the base of the greenhouse is a rectangle 8 ft by 10 ft. The plan shows the base, see Figure 5(b), and the three roof lines, one central for the peak of the roof, and two others running parallel to this for the angle between the sloping sides and the roof. These will be 3 ft from the central roof line on either side.

The area of the front and rear of the greenhouse may be found exactly by splitting the pentagonal shape into a square, 6 ft × 6 ft, two right-angled triangles, 6 ft × 1 ft, and two right-angled triangles, 3 ft × 1.5 ft. Total area

Nets and Surface Area 66

(front and rear) $= 2(36 + 6 + 4.5)$ ft^2 $= 93$ ft^2. The two sides and roof together make a rectangle shape of length 10 ft. The width of this rectangle may be measured from the front elevation as 18.5 ft. Total area of glass is $93 + (10 \times 18.5)$ ft^2 = **278 ft^2**.

SAMPLE QUESTIONS

1. Draw the net of a steel beam of length 6.5 m, with L-shaped cross section shown in Figure 6. Calculate the surface area of the beam to be sandblasted and primed and cost at £4.80 per square metre.

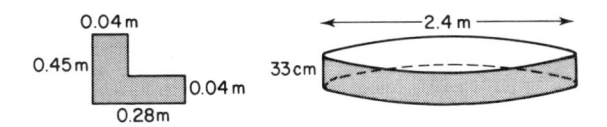

Fig. 6 L-shaped beam   **Fig. 7** Paddling pool

*2. The circular base of the paddling pool shown in Figure 7 is made of thin plastic costing 12p per m^2, and the curved side is made from thicker plastic costing 23p per m^2. Calculate: (*a*) the area of the base, (*b*) the area of the side of the pool, (*c*) the cost of materials for manufacture of the pool. What fraction of the retail price of £4.99 for the pool is for the plastic material?

*3. Calculate the area of the walls of the house drawn in Figure 4. Estimate the cost of cavity wall insulation for this property at £2.50 per square metre of wall filled. Estimate the cost of insulation of the loft area, (not the rear extension) at a cost of £1.20 per square metre.

4. A room is cuboid in shape, length 4.6 m, width 3.8 m, height 3.1 m, with a door, 0.8 m by 2 m, and a window, 2.2 m by 1.8 m. Calculate: (*a*) the area of the ceiling, (*b*) the area of the walls, and (*c*) the area of the floor. Paint for the ceiling costs £4.99 for a tray which covers 9 m^2, paint for the walls costs £3.49 for a tin which covers 18 m^2 and carpet costs £7.99 per m^2. Calculate how many trays and tins of paint must be bought and the total cost of decorating and carpeting the room.

*5. Three toy building blocks are the following shapes; cuboid 6.2 cm by 3.1 cm by 4.6 cm, cylinder radius 2.3 cm, height 6.2 cm, and cone base radius 3.4 cm, slant height 6.2 cm. Calculate which shape has the greatest surface area, and the cost of painting 100 of each block if the paint costs 0.025p per cm^2 covered.

MEASURING
3.9 Volume

The amount of space or matter within a shape is called the **volume** or **capacity** of the shape. A unit of volume is the space in a **cube** having a **standard unit** of length as an edge;

> cubic centimetre 1 cm³ or 1 cc, cubic metre 1 m³,
> cubic decimetre or litre 1 dm³ or 1 l.

Figure 1 shows the filling of a cuboid with unit cubes which gives the basic formula for volume.

> **Volume of a cuboid = length × breadth × height**
> **= area of base × height.**

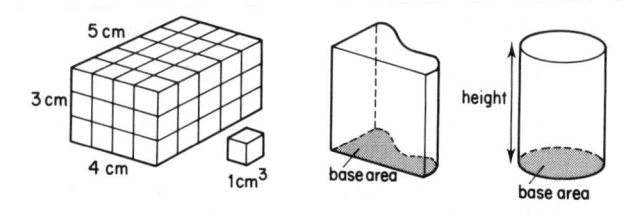

$(3 \times 4 \times 5) \text{cm}^3 = 60 \text{ cm}^3$

Fig. 1 Volume of a Cuboid

(a) General Prism (b) Cylinder

Fig. 2 Volume of Prisms

Prisms (see Solids page 75)

> **Volume of a prism = area of base × height**

Volume of a cylinder = area of base × height = $\pi r^2 h$.

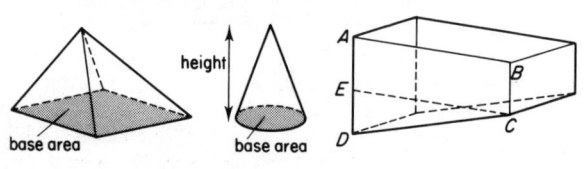

(a) General pyramid (b) Cone

Fig. 3 Volume of Pyramids

Fig. 4 Swimming pool

Volume 68

Pyramids (see Solids page 75)

> *Volume of a pyramid = $\frac{1}{3}$ (area of base × height)

*Volume of a cone = $\frac{1}{3}$ (area of base × height) = $\frac{1}{3}\pi r^2 h$.

Example A rectangular swimming pool, 32 m long, 9.5 m wide, is filled with water which is 2.2 m deep at one end and only 0.8 m at the other. Calculate the volume of water in the pool in (*a*) cubic metres, (*b*) in litres.

Figure 4 shows the shape and dimensions of the water in the pool. The trapezium shaped side *ABCD* of the water forms the base of a prism of height 9.5 m, and once the area of this trapezium is calculated we may use the formula for the volume of a prism given above.
Splitting *ABCD* into a rectangle *ABCE* and a right-angled triangle *CDE*, the area is $(32 \times 0.8) + \frac{1}{2}(32 \times 1.4)\,m^2 = 48\,m^2$.
(*a*) The water has volume $(48 \times 9.5)\,m^3 =$ **456 m^3**.
(*b*) A litre is the volume of a cube with side 10 cm, which is 0.1 m.
 Therefore $1\,l = (0.1 \times 0.1 \times 0.1)\,m^3 = 0.001\,m^3$, and $1000\,l = 1\,m^3$.
 The water has volume $456\,m^3 = 456 \times 1000\,l =$ **456 000 l**.

Example Calculate the volume of a cube of edge 5.25 cm, which has a hole of diameter 3.42 cm drilled directly through the centre of two opposite faces.

Figure 5 shows a diagram of the drilled cube, and its volume can be found by subtracting the volume of the hole from the volume of the solid cube.
A 5.25 cm edge cube has volume $(5.25 \times 5.25 \times 5.25)\,cm^3 = 144.7\,cm^3$ (4SF). The hole is in the shape of a cylinder of height 5.25 cm and radius 1.71 cm. A cylinder is a prism with a circular base, so the volume of a cylinder is $\pi r^2 \times h$ where r is the base radius and h is the height.
The hole has volume $(3.14 \times (1.71)^2 \times 5.25)\,cm^3 = 48.20\,cm^3$ (4SF).
Then the shape has volume $(144.7 - 48.20)\,cm^3 =$ **96.5 cm^3** (3SF).

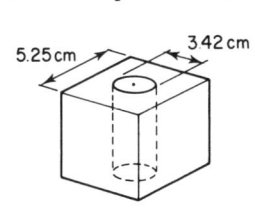

Fig. 5 Drilled cube   **Fig. 6** House roof

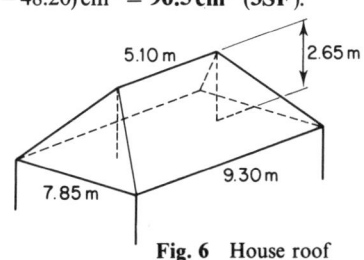

*Example Figure 6 shows the roof of a rectangular house 9.30 m long by 7.85 m wide. The ridge of length 5.10 m is at a height of 2.65 m above the loft floor. Calculate the volume of the roof space. The main walls of the house are 6.30 m high. Calculate the percentage of the complete volume of the house taken up by the roof space.

69 Volume

The roof space can be split into two rectangular based pyramids, $PWXQT$ and $RSZYU$, and a triangular based prism $WXYZUT$ as shown in Figure 7.

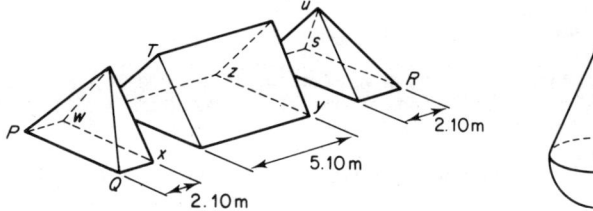

Fig. 7 House roof in sections

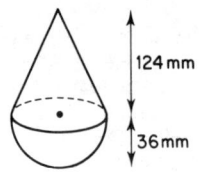

Fig. 8 Wobbling toy

The side QX of the rectangular base of pyramid $PWXQT$ has length $\frac{1}{2}(9.30 - 5.10)\,\text{m} = 2.10\,\text{m}$.
The volume of $PWXQT$ is $\frac{1}{3}(7.85 \times 2.10) \times 2.65\,\text{m}^3 = 14.56\,\text{m}^3$ (4SF).
The triangular end WTX of the prism $WXYZUT$ has area
$\frac{1}{2}(7.85 \times 2.65)\,\text{m}^2 = 10.04\,\text{m}^2$ (4SF),
and so the volume of $WXYZUT$ is $(10.04 \times 5.10)\,\text{m}^3 = 51.20\,\text{m}^3$ (4SF).
The volume of the roof space is $((2 \times 14.56) + 51.20)\,\text{m}^3 = \mathbf{80.3\,m^3}$ (3SF).
The remainder of the house is a cuboid of volume
$(9.30 \times 7.85 \times 6.30)\,\text{m}^3 = 460\,\text{m}^3$,
so the percentage occupied by the roof space is found by expressing $80.3\,\text{m}^3$ as a percentage of the total volume, $(460 + 80.3)\,\text{m}^3 = 540.3\,\text{m}^3$
The percentage is $(80.3 \div 540.3) \times 100 = \mathbf{15\%}$ **(2SF)**.

---

*Example A wobbling clown toy is made of two parts, a hemispherical base of diameter 72 mm and a conical top of height 124 mm, see Figure 8. The base is fabricated in a heavy alloy with density 0.008 g per mm³, and the top is made of a light plastic with density 0.001 g per mm³. Calculate the volume and mass of each part of the toy.

> Volume of a sphere radius r is $\frac{4}{3}\pi r^3$.
> Mass = Volume × density.

The volume of a sphere of diameter 72 mm is $\frac{4}{3}\pi(36)^3\,\text{mm}^3 = 195\,000\,\text{mm}^3$ (3SF), so the hemispherical base will have volume $\mathbf{97\,500\,mm^3}$ **(3SF)**.
The mass of the base $= (97\,500 \times 0.008)\,\text{g} = \mathbf{780\,g}$ **(3SF)**.
The conical top has base radius 36 mm and height 124 mm, so has volume $\frac{1}{3}\pi(36)^2 \times 124\,\text{mm}^3 = \mathbf{168\,000\,mm^3}$ **(3SF)**.
The mass of the top $= (168\,000 \times 0.001)\,\text{g} = \mathbf{168\,g}$ **(3SF)**.

Volume 70

SAMPLE QUESTIONS
1. The four cargo holds of a ship are each shaped like a cuboid, 42 m long, 21 m wide and 37 m deep. Calculate the total volume of space in the four holds.
2. Calculate the volume of cylindrical cans of the following sizes:
   (a) radius 3.2 cm, height 4.5 cm,   (b) diameter 69 mm, height 37 mm.
3. Quarter kilogram packs of butter measuring 11 cm by 7 cm by 5 cm are packed into a case measuring 33 cm by 28 cm by 25 cm. Calculate (a) the volume of a pack of butter,   (b) the volume of the case,   (c) the number of packs of butter in a case. Draw a diagram showing how these packs will fit into the case without any spaces left.
*4 Calculate the cross-sectional area of a pipe of internal radius 0.0035 m. Oil is flowing down the pipe at a speed of 3 ms^{-1} in order to fill a tank of capacity 1.4 m^3. Calculate the volume of oil flowing into the tank in 1 second, and the time taken to fill the tank.
*5 A sweet carton of height 6.2 cm has a base in the shape of a regular hexagon of edge 4.6 cm. Calculate: (a) the area of the hexagonal base,   (b) the volume of the sweet carton. The sweets to fill this carton are roughly cube shaped of edge 2.3 cm. Calculate the number of sweets which would fill the carton if it was possible to pack them without leaving any spaces. In fact, only 20 sweets fit into each carton. Calculate the volume of space left in the carton around the sweets.
*6 It has been estimated that the amount of air breathed by an average person in one day would fill a spherical balloon 2.5 m in diameter. Calculate this quantity: (a) in cubic metres,   (b) in litres. Calculate the edge of a cube which has the same volume as this quantity.
*7 The mass of one cubic metre of water is one metric tonne. Calculate the mass of: (a) 1 cubic centimetre of water,   (b) 1 litre of water, (c) 1 cubic millimetre of a substance which is twice as dense as water.
*8 The cylindrical drum of an upright automatic washing machine is of height 42 cm, and diameter 38 cm. Calculate, in litres, the volume of water needed for small, medium and large loads if the drum is filled to a depth of 22 cm, 28 cm, and 34 cm, respectively. During one automatic cycle the drum is filled once for washing and 4 times for rinsing. Calculate the quantity of water used in one cycle if the large load is selected.
*9 The tomb of Nantukagan was designed in the shape of a square based pyramid. The edge of the base was 35 tukits, and the height was planned to be 45 tukits. In fact the pyramid was only completed up to a height of 26 tukits. Calculate the volume of material which would have been used if the pyramid had been completed and the actual amount used.

GEOMETRY
4.1 Shape and Symmetry   4

Geometry is used to form a model of the real world by the definition of **points, lines, angles, planar shapes** and **solids**. It helps us to appreciate the **symmetry** of objects, their similarities and differences.

Points, lines, planes and angles
A **point** is used to identify a particular position in space.
A **line** is used to describe the set of points which lie directly between two points in space. A line is sometimes considered to extend beyond both points forever.
A **plane** is a two-dimensional, flat set of points, like a piece of paper.
Two lines in a plane either meet in one point or are **parallel** and so never meet, see Figure 1(a).
Two lines which meet at right-angles are **perpendicular**, see Figure 1(b). An example is **horizontal** (flat), and **vertical** (upright), lines.

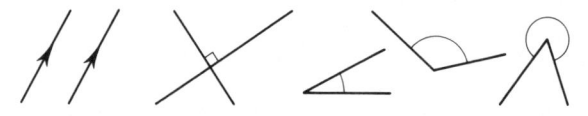

Fig. 1 (a) Parallel (b) Perpendicular lines (c) Acute, Obtuse, Reflex angles.

An **angle** (see Angle, page 51) measures the amount of turning between two lines which meet at a point, see Figure 1(c).
An **acute** (sharp) angle is less than a quarter turn, between 0° and 90°.
An **obtuse** angle is greater than a right angle but less than a half turn, between 90° and 180°.
A **reflex** angle is greater than a half turn, between 180° and 360°.

Symmetry
There are two types of **symmetry**, see Figure 2.
Line symmetry, reflection in a **line**, or **axis of symmetry**.
Rotational symmetry, the symmetry about a point, where the object fits onto itself after rotation through an angle less than 360°.

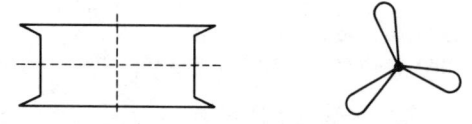

Fig. 2 (a) Line Symmetry (b) Rotational symmetry, order 3.

The **order** of rotational symmetry is the number of such rotations made to turn the shape through 360°, see Figure 2(b).

Shape and Symmetry 72

Example Find the order of rotational symmetry and the number of lines of symmetry for each of the shapes in Figure 2.

The shape in Figure 2(a) has **2** lines of symmetry, shown by dotted lines. If the shape is rotated 180° about its centre it will fit onto itself exactly. Two such rotations will bring the shape back to its original position so its order of rotational symmetry is **2**. The shape in Figure 2(b) has no lines of symmetry, but after a rotation of 120° about its centre the shape fits on top of itself. So the order of rotational symmetry is **3**.

Plane shapes

Triangles

A shape with three sides is called a **triangle**, formed by joining three points by three lines. Triangles are classified by the length of their sides, see Figure 3.
Three equal sides—**equilateral**—3 lines of symmetry, order 3.
Two equal sides and a different length third side—**isosceles**—1 line of symmetry.
All three sides different—**scalene**—no symmetry.

> The sum of the angles of a triangle is 180°. (see page 81 for proof)

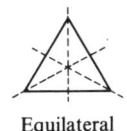

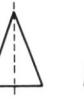

Equilateral   Isosceles   Scalene

Fig. 3 Types of triangle.

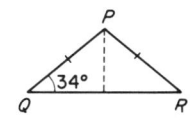

Fig. 4.

Example An isosceles triangle *PQR* has equal sides *PQ* and *PR*, and angle *PQR* is 34°. Calculate the other two angles.

Draw a diagram of the triangle *PQR*, see Figure 4, and then it is clear that angle *PQR* is equal to angle *PRQ*, by the symmetry about a line halfway between *PQ* and *PR*. So angle $PRQ = 34°$.
But the sum of the three angles in triangle *PQR* is 180°, and so,

$\angle PQR + \angle PRQ + \angle QPR = 180°$
$34° + 34° + \angle QPR = 180°$
$\angle QPR = 180° - 68° = \mathbf{112°}.$

Quadrilaterals

A plane shape with four sides is called a **quadrilateral**.

> The sum of the angles of a quadrilateral is 360°

This is proved by splitting the quadrilateral into two triangles.

73 Shape and Symmetry

Quadrilaterals are classified by length of side, size of interior angle and whether opposite sides are parallel, see the table in Figure 5 below.

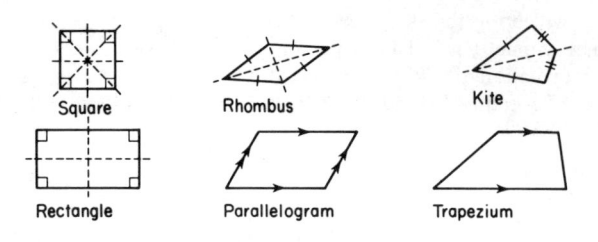

Square   Rhombus   Kite

Rectangle   Parallelogram   Trapezium

	Sq.	Re.	Rh.	Pa.	Ki.	Tr.
Four equal sides	✓		✓			
Two pairs of equal opposite sides	✓	✓	✓	✓		
Two pairs of equal adjacent sides	✓		✓		✓	
Two pairs of parallel sides	✓	✓	✓	✓		
One pair of parallel sides only						✓
Internal angles 90°	✓	✓				
Diagonals bisect each other	✓	✓	✓	✓		
Diagonals meet at right-angles	✓		✓		✓	
Lines of symmetry	4	2	2	0	1	0
Order of rotational symmetry	4	2	2	2	1	1

Fig. 5 Quadrilaterals and their properties.

Example Identify a quadrilateral which is: (*a*) a rhombus with an internal angle of 90°,   (*b*) a trapezium with rotational symmetry order 2.

(*a*) A rhombus has four equal sides, with opposite sides parallel. If one internal angle is 90° then all internal angles must be 90°, by parallel line properties. Therefore the shape is a **square**.

(*b*) A quadrilateral with order of rotational symmetry 2 must have two pairs of opposite sides parallel and equal, since after a rotation of 180° opposite sides must map onto each other. Therefore the trapezium has two pairs of opposite sides equal and parallel and so it is a **parallelogram**.

Polygons
Polygon is a general word used to describe the set of planar *n*-sided shapes, where *n* is any whole number greater than 2.

A **triangle** is a 3-sided polygon, a **pentagon** is a 5-sided polygon,
a **hexagon** is a 6-sided polygon, an **octagon** is an 8-sided polygon.
A **regular** polygon has all its **sides equal** and its **angles equal**.

Shape and Symmetry 74

> * The sum of the interior angles of an n-sided polygon is $(180(n-2))°$.
> The interior angle of an n-sided regular polygon is $(180(n-2)/n)°$.
> The sum of the exterior angles of any polygon is $360°$.

* **Example** Calculate the angle sum, the interior angle and the exterior angle of a regular pentagon.

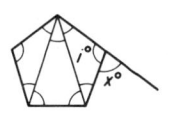

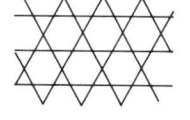

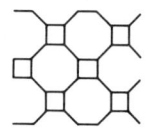

Fig. 6 Pentagon. **Fig. 7** Tessellations.

Figure 6 shows the regular pentagon split into 3 triangles, so the angle sum is $180° \times 3 = \mathbf{540°}$, which is $180(n-2)°$ with $n = 5$.
The interior angle, i, is therefore $540 \div 5 = \mathbf{108°}$, since the five angles must total $540°$. The exterior angle, x, is $180° - 108° = \mathbf{72°}$.

Example Draw a **tessellation** (plane filling pattern) which uses: (*a*) a regular hexagon and an equilateral triangle, (*b*) a regular octagon and a square.

Two possible tessellations are shown in Figure 7. The edges of the polygons used must be the same length if there are to be no gaps left.

SAMPLE QUESTIONS
1 Calculate the number of lines of symmetry and the order of rotational symmetry of a regular n-sided polygon where n is: (*a*) 6, (*b*) 7, (*c*) 8.
2 Find the angles labelled x, y or z in Figure 8. What are the names of the polygons labelled ABC, $JKLM$, $PQRS$ in Figure 8.

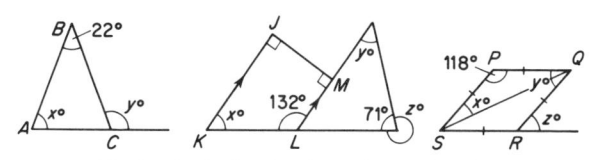

Fig. 8 Angles in triangles and quadrilaterals.

3 Find the interior angle of a regular: (*a*) hexagon, (*b*) triangle, (*c*) octagon, (*d*) dodecagon (12 sides).
4 Draw a tessellation which uses: (*a*) regular hexagons, (*b*) rhombuses, (*c*) kites, (*d*) squares and rhombuses, (*e*) an irregular pentagon.

GEOMETRY
4.2 Solids

Solid Figures
A **solid figure** is a 3 dimensional shape bounded by **faces** (planes), **edges** (lines), and **vertices** (corners or points).

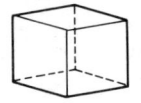

Fig. 1 (a) Cube. 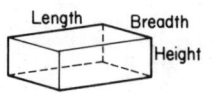 **Fig. 1** (b) Cuboid.

A **cube** has 6 square faces, 12 equal edges and 8 vertices, see Figure 1(a).

A **cuboid** or block has 6 rectangular faces, 12 edges, 8 vertices.

There are three different length edges to a cuboid, usually called the length, breadth and height (Figure 1(b)).

The underneath face of a solid is often called its **base**.

When a solid figure is sliced across by a plane the shape of the planar figure so created is known as a **cross-section**.

Both a cube and a cuboid are examples of a class of solid figures called **prisms**. A prism is '**the same all the way up**', or more precisely, the shape of any cross-section parallel to the base is congruent (identical in size and shape) to the base. A prism is often identified by the shape of its base, so a cuboid is a rectangular based prism, and a cylinder is a circular based prism (Figure 2). One of the most important properties of the prism is that the volume is given by the area of the base multiplied by the height (see Volume, page 67).

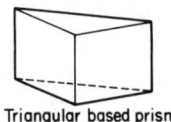

Triangular based prism   I-shape based prism   Cylinder

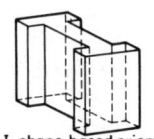

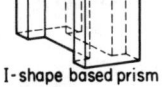

Fig. 2 Prisms.

The other major class of solid figures, those which have a base and all other faces meet at a point at the top, are called **pyramids**.

A pyramid is '**similar all the way up**', or more precisely, the shape of any cross-section parallel to the base is similar (identical in shape, but not in size) to the base. The size of the cross-section is proportional to the distance from the top or point of the pyramid. A pyramid is also identified by the shape of its base. Egyptian tombs are often in the shape of a square based pyramid, and a cone is a circular based pyramid, see Figure 3. The volume of a pyramid is given by one third the area of the base multiplied by the height (see page 67).

Solids 76

Square based Pyramid   Cone

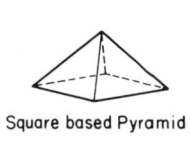

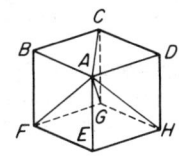

Fig. 3 Pyramids.   **Fig. 4** Dissected cube.

* **Example** A cube *ABCDEFGH*, of edge 3 cm, is cut along planes *AGC*, *AGF* and *AGH* into three solids, see Figure 4. Discribe the shape of the three solids and calculate the volume of each.

The two cuts along planes *AGC* and *AGF* cut off the shape *ABCGF*, **a square based pyramid**, base *BCGF*, height *AB*.
The other two solids are *ADHGC* and *AEFGH*, also identical pyramids.
Therefore the volume of *ABCGF* is $(\frac{1}{3}(3 \times 3) \times 3)$ cm^3 = **9 cm^3**.
This question indicates a method of demonstrating the formula for the volume of a pyramid.

Regular solids

A **regular** solid has faces which are all identical regular polygons. There are five regular solids, called the **Platonic solids**, whose properties are shown in the table and diagrams of Figure 4.

	Shape of face	Number of			Planes of symmetry
		faces	edges	vertices	
Tetrahedron	triangle	4	6	4	6
Cube	square	6	12	8	9
Octahedron	triangle	8	12	6	9
Dodecahedron	pentagon	12	30	20	15
Icosahedron	triangle	20	30	12	15

Fig. 5 The Platonic solids.

SAMPLE QUESTIONS
1. Find the number of faces, edges and vertices of: (*a*) a pentagonal based prism, (*b*) a hexagonal based pyramid, (*c*) an L-shape based prism.
2. Make a model of a cube and clearly mark a cross section in the shape of: (*a*) a square, (*b*) a rectangle, (*c*) a triangle, (*d*) a hexagon.
3. Find the number of planes of symmetry of: (*a*) a cuboid, (*b*) a square based prism, (*c*) a square based pyramid, (*d*) a cone.

GEOMETRY
4.3 Constructions and Loci

4

Ruler and compass constructions

A **ruler** is used to join two points to make a straight line, but it can also be used to measure out a set distance along a line, or to find the distance between two points.

A **pair of compasses** can be used to draw a complete circle, or an arc, of a given radius, from a given centre. In fact compasses are often used to give two arcs of the **same radius**, starting from two different centres, and so locate points which are **equidistant** from the two centres.

Example Draw the set of points equidistant from two points A, B.

Draw two arcs, centre A, with radius just larger than half the length AB. With the **same radius** draw two arcs centre B so as to cut the first arcs. Join the two points where the arcs meet with the line L. Figure 1 shows points A, B, 4 arcs and the line L, the **perpendicular bisector**, or **mediator**, of AB, which is the locus of points equidistant from A and B. The line L and the line AB are at right angles to each other so this construction also gives a way of drawing $90°$ angles.

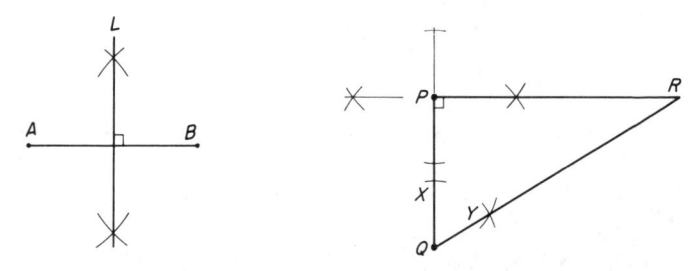

Fig. 1 Finding the set of points equidistant from A and B.

Fig. 2 Constructing a right angle and a $60°$ angle.

Example Draw the triangle PQR where; (a) $PQ = 3.5$ cm, the angle at P is $90°$, the angle at Q is $60°$, (b) $PQ = 5.1$ cm, $PR = 3.6$ cm, $QR = 2.8$ cm.

(a) Draw the side PQ of length 3.5 cm, and extend the line beyond P.
Construct a right angle at P, see Figure 2, as follows:
Draw two arcs, centre P, with same radius, to cut the line PQ either side of P. Construct the perpendicular bisector of these two points using the construction given above, and this will give a right angle at P.
Construct a $60°$ angle at Q as follows:
Draw two arcs, centre Q, with the same radius, one cutting PQ at X and one at about $60°$ to QP. Keeping the same radius, draw a third arc, centre

Constructions and Loci 78

X, to cut the second one at Y. This produces three equidistant points Q, X and Y which form the vertices of an **equilateral** triangle, and so the angle at Q is $60°$, see Figure 2.

Produce (extend) the two sides of the triangle until they meet at R.

(b) Draw the side PQ of length 5.1 cm.
Set the compasses to draw an arc, radius 3.6 cm centre P.
Set the compasses to draw an arc, radius 2.8 cm centre Q.
Label the point of intersetion of the two arcs R, see Figure 3.

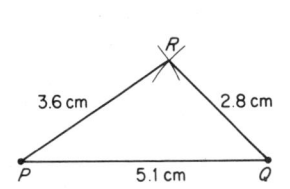

Fig. 3 Constructing a triangle given all three sides.

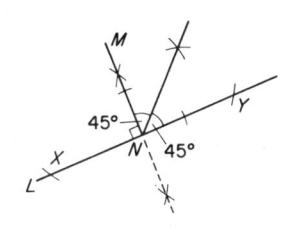

Fig. 4 Constructing the perpendicular to a line L from a point M.

* **Example** Construct a perpendicular from a point M to the line L, meeting L at N, and bisect one of the $90°$ angles at N.

Construct the perpendicular from M to line L as follows:
With centre M, draw an arc which cuts the line L at two points X and Y, see Figure 4. The perpendicular bisector of X and Y will be the required perpendicular from M meeting the line L at N.

Bisect the angle at N as follows:
Draw arcs centre N to cut the line L and the line MN. Draw arcs with the same radius, centred at these two points meeting as shown in Figure 4. Join up N to the intersection of the two arcs to **bisect** the angle of $90°$ at N, giving two $45°$ angles.

Locus

A **locus** is a set of points satisfying the given condition.
Set notation is often used to describe a locus. For example:
The locus $\{P: PA = PB\}$ reads 'the set of points P where the length PA is equal to the length PB'.
If the point P is constrained to move in two dimensions, on a plane, then this locus is the perpendicular bisector of the line AB.
If P is allowed to move in 3 dimensions then the locus is the plane perpendicular to the line AB, midway between the points A and B.

79 Constructions and Loci

Example Figure 5(a) shows a plot of building land beside a railway track. The new main sewer inspection chamber C must be located at a point no further than 50 m from the outfall at O but it must also be at least 30 m away from the railway line L to avoid disturbance. Write these two conditions in set notation and shade the locus of points which satisfy these conditions.

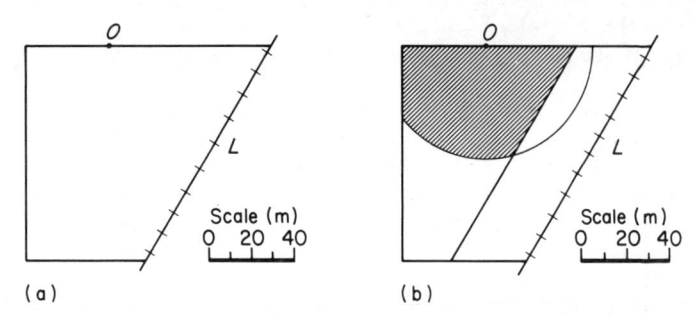

Fig. 5 Location of sewer inspection chamber.

The first condition on the location of C can be written as $\{C: OC \leq 50\}$, and the locus is the inside of a circle, centre O with radius 50 m. The second locus can be written $\{C: CM > 30 \text{ for all } M \in L\}$, and this is a region bounded by a line 30 m away and running parallel to the railway track. The intersection of these two regions is shaded in Figure 5(b).

SAMPLE QUESTIONS

1. Construct the triangle ABC given: (a) $AB = 3.7$ cm, the angle at A is $90°$, $AC = 4.7$ cm, (b) $BC = 88$ mm, $BA = 64$ mm, $CA = 72$ mm, (c) $AC = BC = 4.5$ cm, the angle at C is $45°$.
2. Draw a scalene triangle PQR and construct the perpendicular bisectors of PQ, PR and QR. Using the point where the three bisectors meet as centre draw a circle passing through all three vertices of the triangle PQR.
3. Construct a rhombus with edges of length 7.2 cm and internal angles $30°$ and $150°$. Measure the lengths of the diagonals of the rhombus.
4. Given A and B two fixed points 6 cm apart in a plane, draw the locus of the point P given by: (a) $\{P: AP = 3 \text{ cm}\}$, (b) $\{P: AP > BP\}$, (c) $\{P: \text{angle } ABP = 30°\}$, (d) $\{P: \text{angle } ABP = 90°\}$, (e) $\{P: AP = 5 \text{ cm}, BP < 3 \text{ cm}\}$.

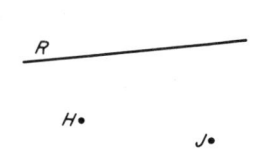

Constructions and Loci 80

Fig. 6 Front paths of houses.

Fig. 7 Radar stations.

5 Figure 6 shows two houses, *H* and *J*, and a road *R*. Copy the diagram and mark the position of a point on the road which is the same distance from both houses. Paths are to be laid from this point to both houses and the boundary line between the two houses is to run from this point perpendicular to the road. Draw the paths and the boundary line.

6 Figure 7 shows a section of coast with two radar stations at *R* and *S* each with a range of 25 miles. An aircraft is approaching the airport directly on a bearing of 060°. Copy Figure 7 and mark the area covered by each radar station and the track of the aircraft. Mark the point on the aircraft's track when Station *S* is first closer than *R*.

7 In Figure 8 a cuboid shaped box is being rolled end over end along a level surface. Draw the locus of the edge marked *C* during one complete rotation of the box.

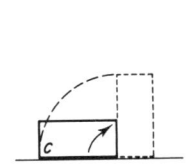

Fig. 8 The rolling box.

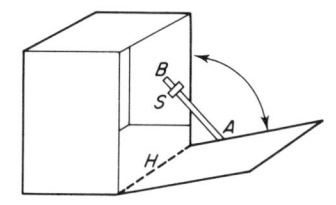

Fig. 9 The drinks cabinet stay.

*8 A metal stay for the pull-down door of a drinks cabinet consists of a straight rod *AB*, jointed to the door at *A*, 13 cm from the hinge *H*, and passing through a sliding joint at *S*, see Figure 9. The position of *S* on the side panel of the cabinet is 3 cm behind and 12 cm above the hinge. The end *B* locks at *S* when the door is opened fully into a horizontal position. Draw an accurate scale drawing of the side view of the cabinet showing the position of the door and stay in the fully open position. Explain why the length of the stay is 20 cm. Now draw the door and stay when the door is inclined at 20°, 40°, 60°, 80° and 90° to the horizontal and draw the locus of the end *B* of the stay as the door closes. What is the minimum depth of the cabinet if the stay is to clear the rear panel?

GEOMETRY
4.4 Prove it
4

Angle theorems
The aim of this section is to demonstrate the idea of **proof**—a logical sequence of steps using **axioms** (starting assumptions) and **theorems** (results) which lead to the desired result. The theorems which follow all involve angles in triangles or circles.

Theorem 1
The sum of the angles of a triangle is 180°. The exterior angle of a triangle equals the sum of the two opposite interior angles.

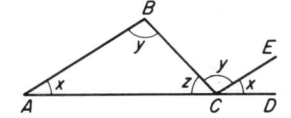

Fig. 1 Angle sum of a triangle.

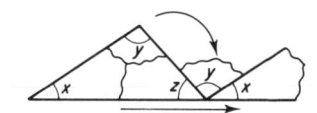

Fig. 2 Paper and scissors proof?

Draw a triangle ABC, shown in Figure 1, with angles x, y and z. Produce the side AC to D. Construct the line CE so that the lines AB and CE are parallel (Figure 1).
$\angle DCE = x$, corresponding angles on a transverse (see page 52)
$\angle BCE = y$, alternate angles on a transverse (see page 52)
So $\angle DCB = x + y$, the exterior angle equals the sum of the two opposite interior angles.
Also $x + y + z = \angle DCE + \angle BCE + \angle ACB = 180°$, since the three angles make a straight line at C, so the angles of a triangle add up to 180°.
Figure 2 shows an alternative, perhaps more familiar, demonstration of this result, which relies on the above proof.
Cut the triangle out of paper marking the three angles, then rip into three parts and reassemble along a line as shown.

Theorem 2
The angle subtended by an arc of a circle at the centre is twice the angle subtended at the circumference.
Draw a circle, centre C, mark a point D on the circumference and join to A and B to give the angles subtended by the arc AB at C and D. Join DC and produce to E, marking angles ADC and BDC as x and y, respectively (Figure 3).
$\angle CAD = x$ and $\angle CBD = y$, since triangles CDA and CDB are isosceles.
$\angle ACE = x + x = 2x$, and $\angle BCE = y + y = 2y$, exterior angles of a triangle.
Then $\angle ACB = 2x + 2y = 2(x + y) = 2 \times \angle ADB$.

The next two results both follow from this theorem.

Prove it 82

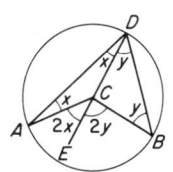

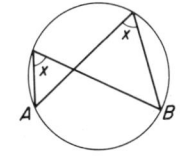

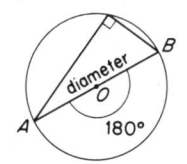

Fig. 3 Angle on an arc.   **Fig. 4** Equal angles.   **Fig. 5** Angle on a diameter.

* **An arc subtends equal angles at the circumference**, since the angle at the centre will be twice any angle at the circumference, see Figure 4.

* **The angle on a diameter is 90°**. The angle at the centre of a semicircular arc is 180°, so any angle at the circumference must be 90°, see Figure 5.

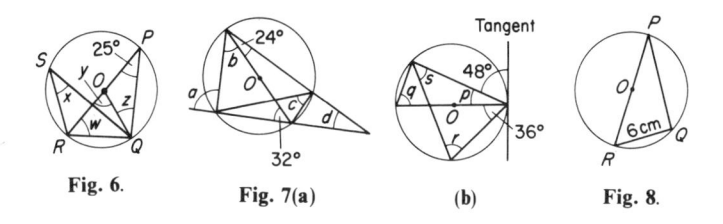

Fig. 6.   **Fig. 7(a)**   **(b)**   **Fig. 8.**

* **Example** Find the angles w, x, y and z marked in Figure 6.

PR is a diameter of the circle centre O, and so $\angle PQR = 90°$.
Since the angles in triangle PQR add up to 180°, $w = \mathbf{65°}$.
Arc RQ subtends both angles $\angle RPQ$ and $\angle RSQ$ at the circumference, so they are equal, and $x = \mathbf{25°}$.
Arc RQ subtends the angle marked y at the centre which must be twice the angle $\angle RPQ$ at the circumference, so $y = 2 \times 25° = \mathbf{50°}$.
Finally, triangle OPQ is isosceles, and so $z = \mathbf{25°}$.

SAMPLE QUESTIONS

* **1** (*a*) Find the angles marked a, b, c and d in Figure 7(a).   (*b*) Find the angles marked p, q, r, and s in Figure 7(b).
* **2** A **cyclic quadrilateral** is a quadrilateral whose 4 vertices lie on a circle. Prove that the opposite angles of a cyclic quadrilateral add up to 180°. (Hint—consider the angles subtended at the centre and circumference by a diagonal of the quadrilateral.)
* **3** Calculate the length of PQ, given that the circle shown in Figure 8 has radius 5 cm.

TRIGONOMETRY
5.1 Trigonometric Ratios

5

The sides and angles of any **right-angled** triangle are connected by three trigonometrical rules and Pythagoras' result. Standard labelling of a right-angled triangle with an angle θ, has three sides o (**Opposite** to θ), a (**Adjacent** to θ), h (**Hypotenuse**, longest side) (Figure 1).

Standard form
$h \sin \theta = o \quad h \cos \theta = a$
$a \tan \theta = o$

Ratio form
$\sin \theta = \dfrac{o}{h} \quad \cos \theta = \dfrac{a}{h} \quad \tan \theta = \dfrac{o}{a}$

Pythagoras' result: $o^2 + a^2 = h^2$

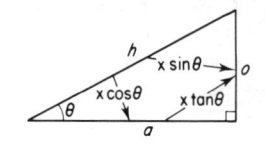

Fig. 1 Basic Trigonometrical results.

Using trigonometry involves **labelling** the diagram, picking the **right rule**, **substituting** the known values and **calculating** the unknown side or angle.

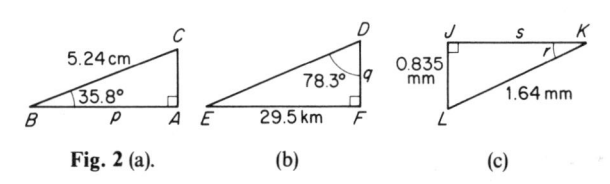

Fig. 2 (a).   (b)   (c)

Example Find p, q, r and s in the triangles in Figure 2.

(a) In triangle ABC we are given angle CBA which will become θ. The side opposite θ, AC, will be labelled (o) and the other sides can now be labelled AB (a) and BC (h). Since we are given BC (h) and we need to find AB (a) we shall use the standard form $h \cos \theta = a$, see Figure 3(a).
$h = 5.24, \quad \theta = 35.8, \quad a = p, \quad h \cos \theta = a$
Substituting for h and θ
$5.24 \cos 35.8 = p$
The value of $\cos 35.8 (= 0.811)$ may be found in mathematical tables or by using a scientific calculator.
Multiplying by 5.24 gives
$p = (5.24 \times 0.8111)$ cm
$= 4.250$ cm $= $ **4.25 cm (3SF)**.

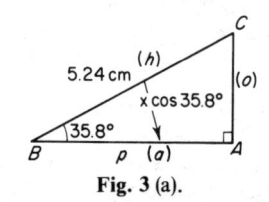

Fig. 3 (a).

(b) In triangle DEF we are given angle EDF which will become θ. The sides can now be labelled EF (o), ED (h) and DF (a). We are given EF (o) and need to find DF (a) so we shall use the standard form $a \tan \theta = o$.

Trigonometric Ratios 84

$o = 29.5$, $\theta = 78.3$, $a = q$, $a \tan \theta = o$
Substituting for o and θ, $q \tan 78.3° = 29.5$
$\tan 78.3° = 4.829$, so $q \times 4.829 = 29.5$
dividing by 4.829, $q = \left(\dfrac{29.5}{4.829}\right)$ km
$q = 6.109$ km = **6.11 km** (3SF).

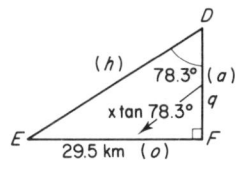

Fig. 3 (b).

(c) In triangle JKL we require angle JKL (labelled r) so the sides will be labelled JL (o), JK (a) and LK (h). We are given JK (o) and KL (h) so we will use the ratio form $\sin \theta = o/h$, see Figure 3(c).

$o = 0.835$, $h = 1.64$, $\theta = r$, $\sin \theta = \dfrac{o}{h}$

Substituting for o and h gives
$\sin r = \dfrac{0.835}{1.64} = 0.5091$.

Using the inverse sine function
on a calculator gives $r = 30.61°$
= **30.6°** (3SF).

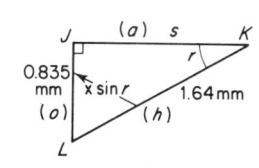

Fig. 3 (c).

To find s we shall use Pythagoras' result, $o^2 + a^2 = h^2$, and substitute for o and h. $0.835^2 + s^2 = 1.64^2$
$0.6972 + s^2 = 2.6896$
Subtracting 0.6972, $s^2 = 1.9924$
Taking square roots, $s = 1.4115$ mm = **1.41 mm** (3SF).

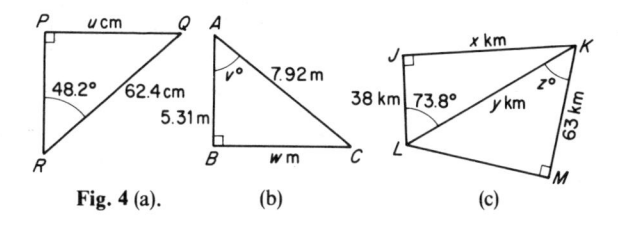

Fig. 4 (a).   (b).   (c).

SAMPLE QUESTIONS

1 Find the values of u, v, w, x, y and z in the triangles of Figure 4.
2 In triangle JKL, JK is 15.4 mm, KL is 23.1 mm and angle $\angle KJL$ is a right angle. Find the length of JL and the size of angles $\angle JLK$ and $\angle JKL$.
3 Find the length of the diagonal AC of the rectangle $ABCD$ with sides AB and BC of length 28.2 cm, 19.5 cm respectively. Calculate the angle between the diagonal AC and the side BC.
4 The isosceles triangle PQR, with PQ and PR equal, can be split into two right-angled triangles PXQ and PXR by joining X, the midpoint of QR, to P. If PR is of length 49.5 km and QR is 31.8 km find the length of PX, the area of the triangle PQR and the angle $\angle PQR$.

TRIGONOMETRY
5.2 Two-dimensional Problems

How to start
1. Draw your own **diagram** of the problem, adding lines to make a right-angled triangle if necessary.
2. Mark on all the **known lengths and angles** and **label** the triangle with o, a and h (as on page 83), relative to one angle θ.
3. Select the right **trigonometrical result**, substitute in the **known values**, and **solve** for the unknown side or angle required.

Example A thin strip of plastic with two holes punched 6.5 cm apart is attached by a pin through one hole to a sheet of squared paper. The pin passes through the origin of a set of axes, scaled in centimetres, drawn on the paper and the plastic strip lies along the x-axis initially (Figure 1). Calculate the coordinates of the other hole if the strip is turned: (a) anticlockwise through 34°, or (b) clockwise through 128°.

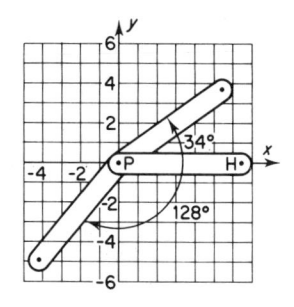

Fig. 1 Plastic strip.   **Fig. 2** (a) 34° turn.   (b) 128° turn.

(a) Draw the plastic strip in the "anticlockwise 34°" position, labelling the pin P, the hole H and the angle 34°, and complete a triangle PHX, where X is the foot of the perpendicular from H to the x-axis, Figure 2(a). Then the coordinates of H are (PX, XH), and we need to find these two sides of the right-angled triangle PHX, by trigonometry. Label the sides (o), (a) and (h), and use $h \cos \theta = a$ to find PX.
6.5 cos 34° = PX, so PX = 6.5 × 0.8290 = 5.389 = 5.4 (2SF)
To find XH, the opposite side, use $h \sin \theta = o$.
6.5 sin 34° = XH, so XH = 6.5 × 0.5592 = 3.635 = 3.6 (2SF)
Then the coordinates of the hole are (**5.4, 3.6**).

(b) Figure 2(b) shows the "clockwise 128°" position, and we use triangle PHY, where Y is the foot of the perpendicular from H to the x-axis. The angle $\angle HPY$ is 180° − 128° = 52°.
Then PY = 6.5 cos 52° = 6.5 × 0.6157 = 4.002 = 4.0 (2SF),
and HY = 6.5 sin 52° = 6.5 × 0.7880 = 5.122 = 5.1 (2SF).
So the coordinates of the hole are (**−4.0, −5.1**), negative because of the position of the hole.

Angle of elevation

Example The **angle of elevation** of the top of a tree is 57°, measured using a clinometer from a point on level ground 25 metres from the base of the trunk. Calculate the height of the tree.

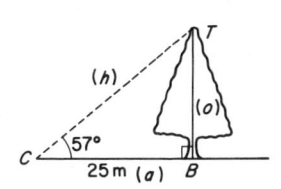

Fig. 3 Height of a tree.

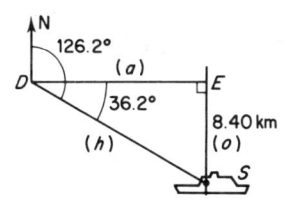

Fig. 4 Bearing of a ship.

Figure 3 shows the tree BT, the clinometer at C and labels the **angle of elevation**, $\angle BCT$, as 57°. The sides of the triangle are labelled (o), (a), (h), relative to this angle, and the known length BC, (a), is marked as 25 m.
The height BT, (o), is required so use $a \tan \theta = o$.
$25 \tan 57° = BT$, so $BT = 25 \times 1.54 = 38.5$,
and the height of the tree is approximately **39 m (2SF)**.

Example A radar station at Dover identifies a ship on a bearing of 126.2°. If the ship is 8.40 kilometres south of the station, find how far east of the station it lies and its direct distance from the station.

Bearings are measured clockwise from North, so draw the station D, the ship S, and mark the bearing 126.2° from North. Draw a due east line from the station and a due north line from the ship, meeting at E, giving an angle $\angle SDE$ which is $126.2° - 90° = 36.2°$, see Figure 4. SE is 8.40 km, and the distances required are DE and DS.
To find DE use $a \tan \theta = o$, giving $DE \tan 36.2° = 8.40$.
Divide both sides by $\tan 36.2°$ ($= 0.732$) gives

$$DE = \frac{8.40}{0.732} = 11.5 \text{ (3SF)}.$$

So the ship is **11.5 km (3SF)** east from the Dover radar station.
To find DS use $h \sin \theta = o$, giving $DS \sin 36.2° = 8.40$.

Divide both sides by $\sin 36.2°$ ($= 0.591$) gives

$$DS = \frac{8.40}{0.591} = 14.2 \text{ (3SF)}.$$

So the ship is **14.2 km (3SF)** from the Dover radar station.

87 Two-dimensional Problems

*Example A scaling ladder is designed to reach the top of a castle wall, 65 feet high, when placed at an angle of 75°, on level ground. Find (*a*) the length of the ladder, (*b*) the distance of the foot of the ladder from the castle wall. When the ladder comes to be used it is discovered that the land slopes away from the castle wall with a gradient of 0.2. Calculate how far the top of the ladder is below the top of the wall when it is placed at 75° to the horizontal.

Figure 5(a) shows the ladder, FT, standing on level ground, reaching the top of the castle wall, BT. FT is inclined at 75° to the horizontal.

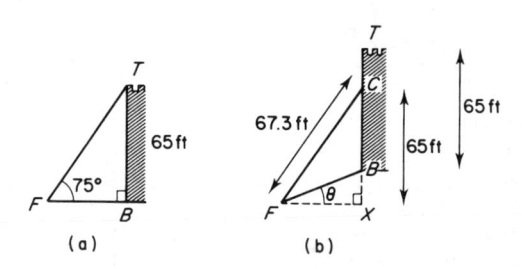

Fig. 5 The ladder and the castle wall.

(a) Using $h \sin \theta = o$ $FT \sin 75° = 65$,
so $FT = 65/(0.966) = 67.3$ (3SF)

(b) Using $h \tan \theta = o$ $FB \tan 75° = 65$,
so $FB = 65/(3.732) = 17.4$ (3SF).
So the ladder is **67.3 ft** long and placed **17.4 ft** from the foot of the wall.

(c) In Figure 5(b) the ladder is shown with its foot at F on the slope, level with the point X directly under the castle wall.
The top rests against the wall at C, and since the ladder is still inclined at 75°, $FX = 17.4$ ft and TC is equal in length to BX.
Now in triangle FBX the angle $\angle BFX$, θ, must satisfy $\tan \theta = 0.2$, since the gradient of the slope FB is $BX/FX = \tan \theta$.
So using $a \tan \theta = o$, $17.4 \tan \theta = BX$, so $BX = 17.4 \times 0.2 = 3.48$.
The top of the ladder is **3.5 ft (2SF)** below the top of the castle wall.

Two-dimensional Problems 88

SAMPLE QUESTIONS
1. The sloping roof of a garage drops 2.5 inches over a distance of 18 feet. Calculate the angle between the sloping roof and the horizontal.
2. A coastguard standing on a cliff top, 34 metres above water level, sights a tanker out at sea when the telescope is inclined at an angle of 2.8° below the horizontal. Calculate the distance of the tanker from the base of the cliff.
3. A balloon used to carry a radio aerial makes the wire slope at an angle of 28° to the vertical when the wind is blowing. If the radio works well so long as the wire reaches a height of 120 m what length of wire is required?
4. A radio mast 43 m tall is supported by 4 wire stays attached to the top of the mast and inclined at an angle of 52° to the horizontal. Find the total length of the 4 wire stays.
5. A radar station spots an unscheduled aircraft at a distance of 22.5 km on a bearing of 225°. Calculate how far west and south the aircraft is from the radar station at this time. Precisely 6 minutes later the aircraft has moved to a position due west at a distance of 18.5 km. Find the average speed of the aircraft during this interval of time.
6. A chimney casts a shadow 72 m long when the sun is sighted at 42° above the horizontal. Calculate the height of the chimney. Some time later the sun is sighted at 53°. Calculate the new length of the chimney's shadow.
*7. A children's slide is inclined at an angle of 41° to the horizontal, see Figure 6. The top of the slide is 3.45 m and the bottom 0.23 m above the ground. Calculate the length of the slide. Find the angle between the step support section and the vertical if it is 3.7 m long.

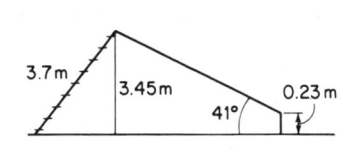

Fig. 6 Children's slide. 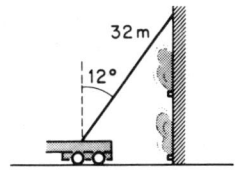   **Fig. 7** Fire-escape ladder.

*8. A fire-escape ladder mounted on a truck 2.2 m above the ground, can be extended of 32 m, so long as it is never inclined at more than 12° to the vertical (Figure 7). Find how close the base of the ladder must be brought to an upright burning building if the full 32 m length is needed. What is the height of the top of the ladder in this position?

TRIGONOMETRY
5.3 Three-Dimensional Problems   5

Use of perspective diagrams
Always start by drawing a large 3-dimensional perspective diagram of the situation, marking any known lengths, angles (especially right angles), compass directions, and bearings.
Label all the vertices, marking the lengths or angles required and construct suitable right-angled triangles to help.
Draw separate 2-dimensional diagrams of each triangle or square needed for further trigonometrical calculations.

*Example An aerial 24.8 m tall is supported by 4 wires, each 32.7 m long, which are joined at the top of the aerial and are fixed to four anchorage points at the corners of a square, see Figure 1. Calculate (*a*) the distance of an anchorage point from the base of the aerial, (*b*) the angle between a wire and the vertical, (*c*) the length of one side of the square formed by the anchorage points.

Fig. 1 (a) Aerial support wires.   (b) Complete labelled diagram.

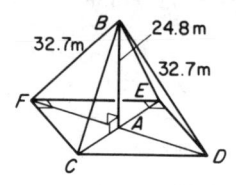

Figure 1(b) shows the aerial AB, the four anchorage points labelled C, D, E, F joined to form the square $CDEF$, where $AB = 24.8$ m, $BC = BD = BE = BF = 32.7$ m.

(*a*) The distance AC can be found by using Pythagoras' result in triangle ABC, drawn separately in Figure 2(a).

$$AC^2 + AB^2 = BC^2$$
Substituting $AC^2 + (24.8)^2 = (32.7)^2$
Squaring $AC^2 + 615.04 = 1069.29$
Take 615.04 $AC^2 = 454.25$
Square root $AC = 21.31$

The distance from an anchorage point to the aerial base is **21.3 m (3SF)**.

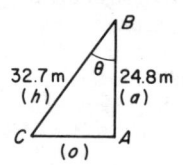

Fig. 2 (a) Triangle ACB.   (b) Square $CDEF$.

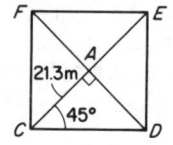

(*b*) The angle between the wire and the vertical is marked θ in Figure 2(a),

Three-dimensional Problems 90

CB is marked h (hypotenuse), BA is marked a (adjacent).
Use the ratio form, $\cos\theta = a/h = 24.8/32.7 = 0.7584$, and so $\theta = 40.67°$.
The angle between the wire and the vertical is **40.7°** (**3SF**).

(c) Square $CDEF$ is drawn in Figure 2(b). $AC = 21.3$ m, angle $CAD = 90°$, so ACD is a right angled triangle, with angle $ACD = 45°$.
Use $h\cos\theta = a$, with hypotenuse CD, adjacent AC and $\theta = 45°$.
$CD\cos 45° = 21.3$, $CD = 21.3/\cos 45° = 21.3/0.707 = 30.12$
The side of the square $= CD =$ **30.1 m** (**3SF**).

*Example The lengths of some of the steel rods for the frame of a playground swing are given in the instruction leaflet, see Figure 3(a), together with dimensions for positioning the rods in a concrete base. Calculate, accurate to 3 significant figures, (a) the angle between the rod marked XY and the base, (b) the length of the rod marked YZ. The seat needs to be 45 cm above the base, but unfortunately the instructions fail to give the length of chain required. (c) Find the total length needed for three seats, each with two supporting chains.

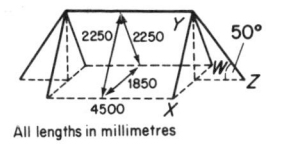

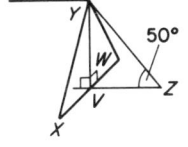

Fig. 3 (a) Playground swing instruction leaflet. (b) Labelled diagram.

Figure 3(b) shows the complete labelled diagram, with V being the point midway between X and W, directly below Y.

(a) Triangle XYW, in Figure 4(a), is isosceles since $YX = YW$, so we split into two right-angled triangles YVX, YVW, where $XV = \frac{1}{2}$ of $XW = 925$ mm.
The angle marked θ may be found by $\cos\theta = a/h$, with adjacent $= XV$, hypotenuse $= YX$. $\cos\theta = 925/2250 = 0.411$, so $\theta = 65.73°$.
The angle between the rod XY and the base is **65.7°** (**3SF**).

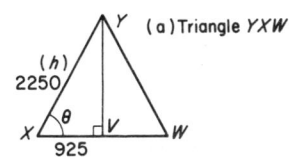

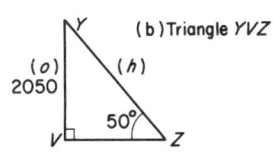

Fig. 4 (a) Triangle YXW,   (b) Triangle YVZ.

(b) To find the length YZ we must first find the side YV of triangle YVX, drawn in Figure 4(a). By Pythagoras' result,

$YV^2 + XV^2 = XY^2$
$YV^2 + 925^2 = 2250^2$
$\quad YV^2 = 5\,062\,500 - 855\,625 = 4\,206\,875$
$\quad YV = 2050$ (3SF)

91 Three-dimensional Problems

Triangle YVZ, drawn in Figure 4(b), has $YV = 2050$, and angle $YZV = 50°$, so we may use $h \sin \theta = o$, where opposite $= YV$ and $\theta = 50°$. $YZ \sin 50° = 2050$, so $YZ = 2050/0.766 = 2676$, and the rod YZ has length **2680 mm (3SF)**.

(c) The length of chain for one side of one seat is $(YV - 450)$ mm $= 1600$ mm.
So for 3 seats we need (6×1600) mm $= 9600$ mm $=$ **10 m to the nearest metre**.

*Example In order to check the height of a lighthouse a surveyor takes sightings from two points P and Q, both at sea level, where P is due south and Q is due east of the lighthouse. The angle of elevation of the top is 22° from P and 37° from Q, Q is on a bearing of 028° from P, and the surveyor walks 76 m between the two points. Find the height of the lighthouse using each of the two angles of elevation, and the average height.

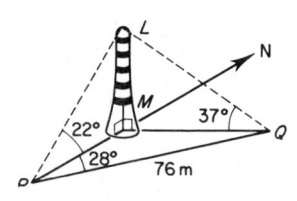

Fig. 5 (a) Lighthouse sightings.   (b) Triangle MPQ.

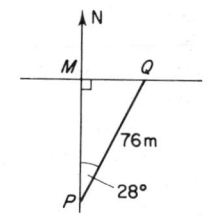

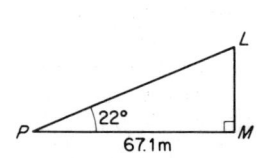

Fig. 5 (c) Triangle PML.   (d) Triangle QML.

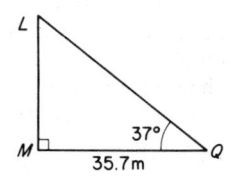

Figure 5(a) shows the lighthouse LM, points P and Q, and the lengths and angles given. Lengths MP and MQ are needed first, so triangle MPQ is drawn in Fig 5(b). Working to 3 significant figures throughout:

$MP = PQ \cos 28° = 76 \times 0.883 = 67.1$,
$MQ = PQ \sin 28° = 76 \times 0.469 = 35.7$

Figure 5(c) and (d) show triangles PML and QML, so:
from P $LM = MP \tan 22° = 67.1 \times 0.404 =$ **27.1 m (3SF)**
from Q $LM = MQ \tan 37° = 35.7 \times 0.754 =$ **26.9 m (3SF)**
The average height given by the two readings is **27.0 m (3SF)**

Three-dimensional Problems 92

SAMPLE QUESTIONS

*1 Mr O'Connor wants to build a car port 8 feet wide on the side of his garage (Figure 6). The garage is 8 ft 6 in high and he has been told to make the roof slope away from the garage at 5° to ensure good runoff for rainwater. Calculate the height of the posts needed to support the lower corners of the car port and the area of glass needed for a car port 15 feet long.

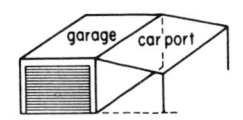

Fig. 6 Car port.

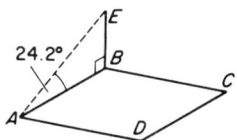

Fig. 7 Parade ground flagpole.

*2 A vertical flagpole 15.8 m high is situated at the corner B of a horizontal square parade ground $ABCD$ (Figure 7). The angle of elevation of the top of the flagpole E from A is 24.2°. Calculate, accurate to 2 significant figures, (a) the side of the parade ground, AB, (b) the length of the diagonal of the square, $ABCD$, (c) the angle of elevation of the top of the flagpole from the centre of the parade ground.

*3 V is the vertex of a right pyramid of height 35 cm and square base $ABCD$. The length of AB is 28 cm. E and F are the midpoints of AB and DC respectively and X is the point of intersection of the diagonals of the square base. Calculate, (a) the angle VFX, (b) the length AX accurate to 1 decimal place, (c) the angle which the edge VA makes with the square base.

*4 The roof of an L-shaped bungalow (see Figure 8) has the dimensions $AB = AC = FE = FD = 4.3$ m, $BC = ED = CG = EG = 6.8$ m, and all the roofs slope at the same angle. Calculate, (a) the roof angle BCA, (b) the height of the apex A above BC, (c) the length GH.

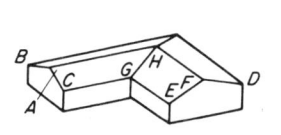

Fig. 8 Bungalow roof.

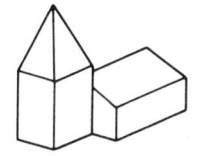

Fig. 9 Church Spire.

*5 A church tower 12.5 m high, is surmounted by a spire in the shape of a square based pyramid (Figure 9). The square base has edges of length 4.2 m, and the sloping faces are inclined at 26° to the vertical. Find, (a) the height of the tip of the spire above the ground, (b) the total area of the four faces of the spire.

TRIGONOMETRY
5.4 Making Waves

5

Sine, cosine, tangent graphs for 0 to 360 degrees

$y = \sin x$, $y = \cos x$ and $y = \tan x$ are the three basic trigonometric functions. They were introduced as ratios between sides of right-angled triangles (see page 83), but they may be evaluated for any x, using a scientific calculator, see Table 1 below.

Table 1 The trigonometric functions.

x	0	30	60	90	120	150	180	210	240	270	300	330	360
$\sin x$	0	0.5	0.87	1	0.87	0.5	0	−0.5	−0.87	−1	−0.87	−0.5	0
$\cos x$	1	0.87	0.5	0	−0.5	−0.87	−1	−0.87	−0.5	0	0.5	0.87	1
$\tan x$	0	0.58	1.73		−1.73	−0.58	0	0.58	1.73		−1.73	−0.58	0

Figure 1 shows these values plotted and joined. The sine and cosine functions are **waves** with **amplitude** 1 and **period** 360°. The tangent function tends to **infinity** as x tends to 90°, so has an **asymptote** at $x = 90, 270, \ldots$, and **period** 180°.

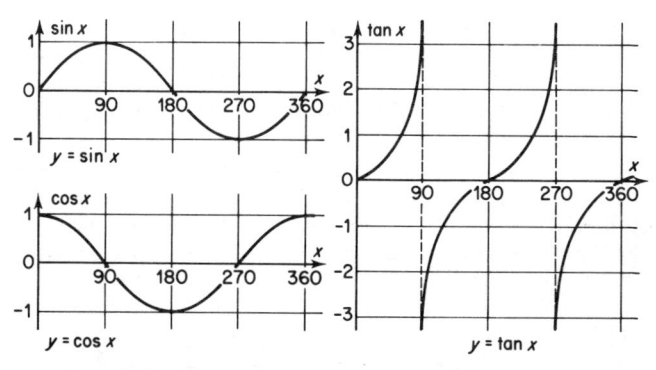

Fig. 1   The trigonometrical graphs.

***Example**   The height above ground level, h, measured in metres, of a seat on a fairground Big Wheel is given by the equation $h = 4 \sin 3t + 5$, where t is the time in seconds. Calculate a table of values for h where t goes from 0 to 120, in 10 second intervals. Plot the graph of h and use your graph to read off;   (a) the maximum height reached by the seat,   (b) the diameter of the Big Wheel,   (c) the length of time for the wheel to make one complete turn,   (d) the speed of rotation of the wheel in revolutions per minute (rev/min).

Use the normal device of breaking down the evaluation of h into easy steps to help calculate the values in Table 2.

Making Waves 94

Table 2

t	0	10	20	30	40	50	60	70	80	90	100	110	120
$3t$	0	30	60	90	120	150	180	210	240	270	300	330	360
$\sin 3t$	0	0.5	0.87	1	0.87	0.5	0	−0.5	−0.87	−1	−0.87	−0.5	0
$4 \sin 3t$	0	2	3.5	4	3.5	2	0	−2	−3.5	−4	−3.5	−2	0
$4 \sin 3t + 5$	5	7	8.5	9	8.5	7	5	3	1.5	1	1.5	3	5

Figure 2 shows the graph of h, plotted for t from 0 to 120.

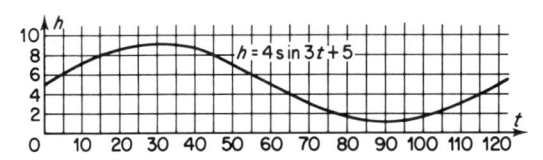

Fig. 2 The height of a seat on a Big Wheel.

(a) The maximum height of the seat is **9 m** when $t = 30$.
(b) The difference between the greatest and least heights of the seat is $(9-1)$ m = **8 m**, which must be the diameter of the Big Wheel.
(c) The wheel makes one turn in 120 seconds = **2 min**.
(d) The speed of rotation is therefore $\frac{1}{2}$ **rev/min**.

SAMPLE QUESTIONS

*1 Calculate a table of values and plot the graph of the functions; (a) $y = 3 \sin x + 1$, $0 \leqslant x \leqslant 360$, (b) $y = 6 - 2 \cos 4x$, $0 \leqslant x \leqslant 90$.

*2 The depth d, measured in feet, of water at a seaside harbour, varies with time according to the formula $d = 5 \cos 15t + 8$, where t is measured in hours after midnight. Calculate a table of values for d during one 24 hour day and plot the graph of d. Use your graph to find, (a) the depth of water at 0800 hrs, (b) the time when the water is at its least depth, (c) the period of time during the day when a boat which needs water at least 6 feet deep can enter the harbour.

*3 A vertical spring has a small weight attached to one end and is oscillating up and down. The length, l cm, of the spring at time t s, is given as $l = 18 \sin 200t + 34$. Calculate the length of the spring for $t = 0$ to 3 s in intervals of 0.2 s. Draw a graph of l and use your graph to find, (a) the greatest length of the spring during the following motion, (b) the first time when the spring is at its minimum length, (c) the time for one complete oscillation of the weight.

EQUATIONS AND GRAPHS
6.1 Algebra

In **algebra** we use letters to represent numbers, so that x may stand for any number. The letter x is called a **variable**.
These variables are combined into;

	Example	Meaning
terms	$3x^2$	'x squared times 3'
expressions	$5y - (2+z)$	'5 times y minus the expression (2 plus z)'
formulae	$v = u + at$	'v equals u plus a times t'

The usual laws of operations on numbers apply, brackets first, then powers, multiplication and division, finally addition and subtraction.

Substitution

The **value** of an algebraic expression may be found by **substituting** numbers for the variables and calculating the result.

Example Given $a = 2$, $b = 4$, $c = 12$ and $d = -3$ evaluate:

(a) ab (b) $ad + bc$ (c) $d + \sqrt{b}$ (d) $3b^2$ (e) $\dfrac{c}{d}$ (f) $\dfrac{d^2 - 5b}{4a}$

Brackets have been inserted to show which operation is to be done first.

(a) $ab = 2 \times 4 = \mathbf{8}$

(b) $ad + bc = (2 \times -3) + (4 \times 12)$
$= -6 + 48 = \mathbf{42}$

(c) $d + \sqrt{b} = -3 + (\sqrt{4})$
$= -3 + 2 = \mathbf{-1}$

(d) $3b^2 = 3 \times (4^2)$
$= 3 \times 16 = \mathbf{48}$

(e) $\dfrac{c}{d} = \dfrac{12}{-3}$
$= 12 \div (-3) = \mathbf{-4}$

(f) $\dfrac{d^2 - 5b}{4a} = \dfrac{(-3)^2 - (5 \times 4)}{4 \times 2}$
$= \dfrac{9 - 20}{8} = \dfrac{\mathbf{-11}}{\mathbf{8}}$

Example Evaluate the following when $p = 7$, $q = 4$, $r = 9$ and $s = -5$:

(a) $(p-q)^2$, (b) $(q - 2r)(p - s)$, (c) $3 - (2p - q)/s$.

(a) Find the value of the bracket before squaring:
$(p - q)^2 = (7 - 4)^2 = 3^2 = \mathbf{9}$.

(b) Evaluate each bracket first before multiplying them together:
$(q - 2r)(p - s) = (4 - (2 \times 9)) \times (7 - (-5)) = (4 - 18) \times (7 + 5)$
$= -14 \times 12 = \mathbf{-168}$.

(c) The division by s must be carried out before subtraction from 3:
$3 - (2p - q)/s = 3 - ((2 \times 7) - 4)/(-5) = 3 - (10)/(-5)$
$= 3 - (-2) = \mathbf{5}$.

Algebra

Expanding products and simplifying

When **expanding** expressions which contain a product involving brackets, the contents of the bracket are each multiplied by the term outside, and the expression is then **simplified** by collecting **like terms**.

Example Expand and simplify the following:
(a) $3(x+4) + 4(3x-2)$ (b) $7x - 3(2x - 5y)$ (c) $4a(a+b-c)$
*(d) $(a+2b)(a-3b)$ *(e) $\dfrac{(2x-1)}{-3} - \dfrac{(4-x)}{2}$.

(a) $\;\;3(x+4) + 4(3x-2)$ \qquad (b) $\;\;7x - 3(2x-5y)$
$\;\;\;\;= 3x + 12 + 12x - 8$ \qquad\quad $\;\;\;\;= 7x - 6x + 15y$
$\;\;\;\;= \mathbf{15x + 4}$ \qquad\qquad\qquad\quad $\;\;\;\;= \mathbf{x + 15y}$

(c) $\;\;4a(a+b-c)$ \qquad\qquad (d) $\;\;(a+2b)(a-3b)$
$\;\;\;\;= \mathbf{4a^2 + 4ab - 4ac}$ \qquad\quad $\;\;\;\;= a(a-3b) + 2b(a-3b)$
$\;= a^2 - 3ab + 2ba - 6b^2$
$\;= \mathbf{a^2 - ab - 6b^2}$

(e) $\dfrac{(2x-1)}{3} - \dfrac{(4-x)}{2} = \dfrac{2(2x-1) - 3(4-x)}{6} = \dfrac{4x - 2 - 12 + 3x}{6}$

$\qquad\qquad\qquad = \dfrac{7x - 14}{6} = \mathbf{\dfrac{7(x-2)}{6}}$

SAMPLE QUESTIONS

1 Given $a = 1$, $b = -2$, $c = 3$ and $d = -4$, evaluate:
(a) $a+b$ (b) $b+c$ (c) $a-d$ (d) $b-c$
(e) $5b$ (f) $2c$ (g) $5b+2c$ (h) $5a-2d$
(i) $4a+3b+2c+d$ (j) bc (k) $a+bd$ (l) b^2+bc
(m) $bc+cd$ (n) $\dfrac{b}{c}$ (o) $\dfrac{b+d}{c}$ (p) $\dfrac{a-b}{b-c}$

2 Given $p = 7$, $q = -5$, $r = -12$, $s = 2$, evaluate:
(a) $2r+3s$ (b) $pq-rs$ (c) $q+s^2$
(d) $3p+qs$ (e) $(p-q)/r$ (f) $p(qr-p)$

3 Given $w = -3$, $x = -7$, $y = 7$ and $z = -5$, evaluate:
(a) $x(w-2z)$ (b) $3y(2y-5z)$ (c) $x(5+2wy) - y(8-xz)$

4 Expand and simplify:
(a) $3(2x+y) + 4(3x+2y)$ (b) $2(3a+4b) - 3(5a-3b)$
*(c) $7a(2a+b) - 3(a^2+4ab)$ *(d) $6mn - m^2 - 4m(n-m)$

*5 Simplify: (a) $\dfrac{(3x+2)}{2} + \dfrac{(2x-4)}{3}$ (b) $\dfrac{(5x-1)}{3} + \dfrac{(2-x)}{5}$

EQUATIONS AND GRAPHS
6.2 Factorisation

6

The opposite process to multiplying out brackets is **factorisation**, writing the expression as a **product** of terms.

Example Factorise: (a) $6a + 2ab - 3a + ab$, (b) $8xyz - 12xy$, (c) $3t^2 - 14t + 6ts$.

(a) Collect **like terms**. Spot the **common factor** $3a$, and factor it out from each term. Write the expression as a product, $3a$ times a bracket which contains the sum of products from each term.

$$6a + 2ab - 3a + ab = 3a + 3ab = 3a(1) + 3a(b) = \mathbf{3a(1 + b)}.$$

(b) The **common factor** is $4xy$.
$$8xyz - 12xy = 4xy(2z) - 4xy(3) = \mathbf{4xy(2z - 3)}.$$

(c) $3t^2 = 3 \times t \times t$, so the only **common factor** is t.
$$3t^2 - 14t + 6ts = t(3t) - t(14) + t(6s) = \mathbf{t(3t - 14 + 6s)}.$$

*Quadratic factorisation

When two linear brackets, $(x + 3)$ and $(x + 2)$, are multiplied together the result is a **quadratic** expression. Such expressions are harder to factorise, and understanding the multiplication of brackets is a good preparation to factorising quadratics.

$$(x + 3)(x + 2) = x \times x + 3 \times 2 + x \times 2 + 3 \times x$$
$$= x^2 + x(3 + 2) + (3 \times 2) = x^2 + 5x + 6.$$

The loops show which terms have been multiplied together. The two 'eyebrows' give the x^2 and constant terms, whilst the 'mouth and nose' give the two x terms, $3x$ and $2x$.

So, in general, $(x + p)(x + q) = x^2 + ax + b$, if $p + q = a$ and $pq = b$.

Example Factorise (a) $x^2 + 8x + 15$, (b) $x^2 - 6x + 8$, (c) $x^2 - x - 12$.

(a) Using the notation above, we need p and q so that $p + q = 8$ and $pq = 15$. The factors of 15 are 1×15 or 3×5, and $3 + 5 = 8$, so we try $(x + 3)(x + 5)$.

$$\mathbf{(x + 3)(x + 5)} = x^2 + 15 + 5x + 3x = x^2 + 8x + 15$$

(b) We need $p + q = -6$ and $pq = 8$. The negative factors of 8 are $(-1) \times (-8)$ or $(-2) \times (-4)$, and $-2 + -4 = -6$, so we try $(x + -2)(x + -4) = (x - 2)(x - 4)$.

$$\mathbf{(x - 2)(x - 4)} = x^2 + 8 + -2x + -4x = x^2 - 6x + 8$$

(c) We need $p + q = -1$ and $pq = -12$. The factors of -12 are $(-1) \times (12)$, $(-2) \times (6)$, $(-3) \times (4)$, $(-4) \times (3)$, $(-6) \times (2)$ and

Factorisation

$(-12) \times (1)$, but the only pair adding to -1 is -4 and 3, so we try $(x-4)(x+3)$.

$(x-4)(x+3) = x^2 - 12 - 4x + 3x = x^2 - x - 12$

When the quadratic starts with a term like $3x^2$, we must use a more general factorisation formula.

$(px+q)(rx+s) = prx^2 + qs + psx + qrx = prx^2 + x(ps+qr) + qs$

Example Factorise: (a) $2x^2 + 7x + 3$, (b) $6x^2 - 11x - 10$ (c) $4x^2 - 9$.

(a) The $2x^2$ term must come from $(2x) \times (x)$ and the 3 from 1×3. So there are two possible sets of factors, $(2x+3)(x+1)$, $(2x+1)(x+3)$, avoiding negative numbers initially. The second pair works.

$(2x+1)(x+3) = 2x^2 + 3 + 6x + x = 2x^2 + 7x + 3$

(b) The $6x^2$ term might come from $(x) \times (6x)$ or $(2x) \times (3x)$ and the -10 from $(-1) \times (10), (-2) \times (5), (-5) \times (2)$ or $(-10) \times (1)$. Therefore there are 16 possible pairs of factors, the first two being $(x-1)(6x+10)$ and $(6x-1)(x+10)$. The factors which fit the quadratic are $(2x-5)$ and $(3x+2)$.

$(2x-5)(3x+2) = 6x^2 - 10 + 4x - 15x = 6x^2 - 11x - 10$

(c) This is an example of the **difference of 2 squares**, since both terms are perfect squares, $4x^2 = 2x \times 2x$, and $9 = 3 \times 3$. The factors of this quadratic are $(2x+3)$ and $(2x-3)$.

$(2x+3)(2x-3) = 4x^2 - 9 + 6x - 6x = 4x^2 - 9$

Once this process becomes more automatic it is possible to eliminate trying *all* the pairs of factors, thus focusing on the correct pair more quickly.

SAMPLE QUESTIONS

1. Factorise: (a) $2pq - p$ (b) $12q^2 + 6pq - 3p$
   (c) $3r + 2s - r + 4s$ (d) $3x^2 y - 6xy + 9xy^2$
   (e) $5a - 10b^2 + 20a^2$ (f) $56abc - 42b^2 c$

*2 Complete these quadratic factorisations:
   (a) $x^2 - 3x - 28 = (x+4)(x\ \ )$ (b) $x^2 + x - 6 = (x\ \ 2)(x\ \ )$
   (c) $2x^2 - 11x - 40 = (2x\ \ )(x\ \ 8)$ (d) $4x^2 - 9 = (2x\ \ )(2x\ \ )$

*3 Factorise: (a) $x^2 + 6x + 9$ (b) $x^2 - 7x + 12$ (c) $x^2 + x - 12$
   (d) $x^2 - 2x - 15$ (e) $2x^2 + 11x + 5$ (f) $3x^2 - 13x + 4$
   (g) $2x^2 - x - 3$ (h) $4x^2 + 5x - 6$ (i) $6x^2 - 23x - 18$

EQUATIONS AND GRAPHS
6.3 What is a Function? 6

A **function** f is a **mapping** from one set to another, using a carefully defined rule or formula, so that each element, x, of the first set is mapped to a particular element, $f(x)$, of the second set. Functions are often mappings from one set of numbers to another, as follows:

$f: x \rightarrow 3x + 1$, which reads 'f is the function which **maps** x to $3x + 1$'.
Sometimes a function is written in **equation** form, $y = 3x + 1$. Then, for any value of x, the corresponding value of y is found from the equation $y = f(x) = 3x + 1$.

Mapping diagram and table of values

A **mapping diagram** shows how the function may be broken down into steps which demonstrate which **operations** are being carried out (Figure 1). In this case the two operations are 'multiply by 3' and 'add 1'.

	×3	+1				x	y
f: x	→ $3x$	→ $3x+1$					
2	→ 6	→ 7	$f(2)$	= 7		2	7
3	→ 9	→ 10	$f(3)$	= 10		3	10
7	→ 21	→ 22	$f(7)$	= 22		7	22
−5	→ −15	→ −14	$f(-5)$	= −14		−5	−14

Fig. 1 Mapping diagram and table of values for the function
f: $x \rightarrow 3x + 1$, or $y = f(x) = 3x + 1$.

The **table of values** for a function is a set of **coordinate pairs** (x, y), where the value of y in the pair corresponds to the value of x. Figure 1 shows how such pairs can be found and tabulated.
So if $x = 2$, then $y = 3 \times 2 + 1 = 7$, giving the pair $(2, 7)$.

Cartesian graphs

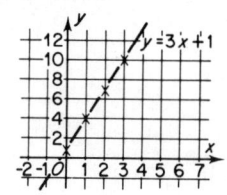

Fig. 2 The graph of the function f, $y = f(x) = 3x + 1$

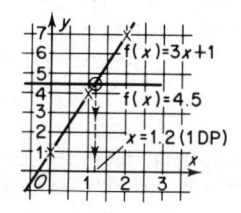

Fig. 3 Solving $3x + 1 = 4.5$

One of the easiest ways of understanding a function is from its **Cartesian graph**. This uses two axes Ox and Oy at right angles, and the coordinate pairs (x, y), calculated earlier for the function f, are plotted as points and joined to give the **graph of the function**. In the example above the points lie on a straight line, see Figure 2, with equation $y = 3x + 1$.

Graphical solution of equations

The graph of a function f may be used to solve an equation of the form $f(x) = c$, where c is a constant.

Example Solve the equation $3x + 1 = 4.5$ graphically.

The left hand side of this equation is the function $f(x) = 3x + 1$, already drawn in Figure 2 above. Draw the graph of the **constant** function $f(x) = 4.5$ on the same graph (Figure 3).

The two lines meet at the point (1.2, 4.5), approximately, and the solution of the equation $3x + 1 = 4.5$ is therefore $x = $ **1.2 (1DP)**

Example A mathematically inclined foreman has noted that when he puts x men to digging a trench it takes them $60/x$ minutes to finish, so long as he uses no more than 6 men. If he uses more it still takes 10 minutes. Find a function to describe this and draw its graph. How many men should he use if he wants the job finished in, (a) 15, (b) 5, minutes?

The function $f: x \longrightarrow \begin{cases} 60/x, & \text{if } x < 6 \\ 10, & \text{if } x \geqslant 6 \end{cases}$ describes this situation.

x	1	2	3	6	8
f(x)	60	30	20	10	10

Fig. 4 Table of values and Cartesian graph of the foreman's problem

Figure 4 shows a table of values and the graph of this function. To finish the trench in 15 minutes needs **4 men**, by following the dotted line on the graph. Clearly it is **impossible** to finish in 5 minutes!

SAMPLE QUESTIONS
1. The function f, is given by $f: x \to 9 - 2x$. Find: (a) $f(1)$, (b) $f(4)$, (c) $f(7)$, (d) $f(-3)$, (e) x such that $f(x) = -1$.
2. The function f is represented by the mapping diagram

   $x \xrightarrow{\times 2} \boxed{?} \xrightarrow{+1.8} \boxed{?} \xrightarrow{\text{square}} f(x)$.

   Copy and complete the mapping diagram and find: (a) $f(2.3)$, (b) the function $f(x)$. Solve the equation $f(x) = 0$.
3. $f(x) = (5 + 2x)/3$. Find: (a) $f(-3)$, (b) x such that $f(x) = 12$.
4. Plot the coordinate pairs of f, given in the table, and join with a smooth curve. Use your graph to solve the equation $f(x) = 5$.

x	−3	−2	−1	0	1	2	3	4	5	6	7
f(x)	2.8	5.0	7.0	8.4	9.2	9.4	9.3	8.5	7.3	5.6	3.5

EQUATIONS AND GRAPHS
6.4 Conversion Graphs

6

A **conversion graph** is used to display the relationship between two quantities, often connected by a formula. Given the value of one quantity the graph is used to read off the corresponding value of the other quantity.

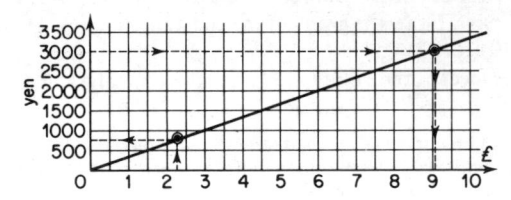

Fig. 1 Currency conversion graph, pounds sterling and Japanese yen

Example The conversion graph drawn in Figure 1 may be used to convert between pounds sterling and Japanese yen. From the graph find: (a) £2.30 in yen, (b) 3000 yen in pounds sterling, (c) the current exchange rate (the number of yen which may be exchanged for £1).

(a) Draw a vertical line up from £2.30 on the pounds sterling axis until it meets the conversion line, then from the point of intersection draw a horizontal line across to the Japanese yen axis.
Reading off the amount shows that £2.30 can be exchanged for **750 yen**.
(b) Reverse this procedure for 3000 yen gives an exchange value of **£9.10**.
(c) £10 can be exchanged for 3300 yen, so £1 can be exchanged for **330 yen**.

Example The function $f: C \to F$, where $F = 1.8C + 32$, converts a temperature C in degrees Celsius into the corresponding temperature F in degrees Fahrenheit. Calculate a table of values of F for values of C from -10 to 110 at intervals of 20. Plot these values on a graph using 1 cm for 10 degrees C and 1 cm for 20 degrees F, and join with a straight line. Use the graph to find: (a) 25°C in degrees Fahrenheit, (b) 155°F in degrees Celsius, (c) boiling point of water, 100°C in degrees Fahrenheit.

Use the two operations 'multiply by 1.8' and 'add 32' on the value of C to find the corresponding value of F, shown in the table below.

C	-10	10	30	50	70	90	110	↓ × 1.8
1.8C	-18	18	54	90	126	162	198	↓ + 32
F	14	50	86	122	158	194	230	

Plot the points, using the defined axes, and join to give the graph shown in Figure 2, scaled down to fit this page.

Conversion Graphs 102

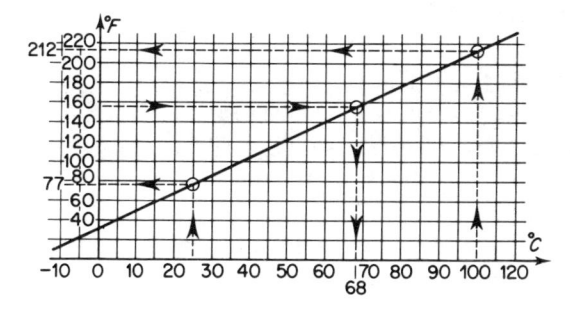

Fig. 2 Conversion graph for temperature °Celsius to °Fahrenheit

Reading from the graph (a) 25° Celsius corresponds to **77° Fahrenheit**, (b) 155° Fahrenheit corresponds to **68° Celsius**, (c) 100° Celsius corresponds to **212° Fahrenheit**, the boiling point of water.

SAMPLE QUESTIONS

1 The table of values shows the cost C in pounds charged by the Gas Board for using G hundreds of cubic feet of gas in one quarter.

G 100 ft^3	13	39	91
C £	14.20	24.20	44.20

Draw a straight line graph through the points plotted from these figures using a scale of 1 cm to 100 cubic feet of gas and 2 cm for £1 cost. Use your graph to find: (a) the cost of using 5000 cubic feet of gas in the quarter, (b) the amount of gas used if the cost is £20.00, (c) the standing charge (made if no gas is used).

2 The quarterly rental of a telephone line is £15 and the cost of calls 5.75p per unit. Calculate the charge for using; (a) 0, (b) 200, (c) 400, units in one quarter. Use these figures to draw a graph of the charges for using between 0 and 500 units in the quarter, with a scale of 2 cm for 100 units and 2 cm for £10. Use your graph to find: (d) the charge for using 85 units in the quarter, (e) the number of units used if the charge is £35.

3 A service engineer charges £25 for a call-out which includes the first hours work, and £15 per hour for work done after the first hour. Write down the cost for a call-out lasting; (a) ½ an hour, (b) 1 hour, (c) 2 hours, (d) 5 hours. Draw a graph, consisting of two lines to show the charges for call-outs lasting from 0 to 5 hours. Use your graph to find; (e) the charge for a 90 minute call-out, (f) the length of call-out for which the charge is £75.

EQUATIONS AND GRAPHS
6.5 Travel Graphs

6

Distance, speed and time
When travelling at a steady or **constant speed** of S kph (kilometres per hour), the **distance** travelled D km, in a **time** T hours, is given by the formula $D = ST$, | distance = speed × time |.

This formula can be rearranged in two ways $S = D/T$ and $T = D/S$. When the speed changes the formula gives the **average speed**.

$$\text{average speed} = \frac{\text{distance travelled}}{\text{time taken}}$$

Example A man walks at a steady speed of 5 km/h. (*a*) How far does he walk in 3 hours? (*b*) How long will he take to cover the first 2 kilometres? (*c*) After 3 hours he slows down and covers the last kilometres in 20 minutes. What is his new speed? (*d*) What is his average speed over the whole journey?

(*a*) Distance = speed × time = (5 km/h) × (3 h) = **15 km**.
(*b*) Time = distance/speed = (2 km)/(5 kph) = 0.4 h = **24 minutes**.
(*c*) Speed = distance/time = 1 km in 20 min = 3 km in 1 hour = **3 km/h**.
(*d*) Total distance travelled is (15 + 1) km = 16 km.
   Time taken is 3 hours 20 minutes = $3\frac{1}{3}$ h.
   Average speed = total distance/time = (16 km)/($3\frac{1}{3}$ h) = **4.8 km/h**.

SAMPLE QUESTIONS
1. Find the average speed for the following journeys:
   (*a*) 120 km in 4 h,   (*b*) 714 km in 7 h,
   (*c*) 50 km in $\frac{1}{2}$ h,   (*d*) 15 km in $2\frac{1}{2}$ h,
   (*e*) 10 cm in 20 min,   (*f*) 252 miles in 4 h 12 min.
2. How long will it take a car to travel:
   (*a*) 300 km at 50 km/h,   (*b*) 60 km at 120 km/h,
   (*c*) 400 miles at 70 mile/h,   (*d*) 5 miles at 25 mile/h,
   (*e*) 2.8 km at 56 km/h,   (*f*) $5\frac{1}{2}$ miles at 30 mile/h.
3. How far does a plane travel if it flies at an average speed of:
   (*a*) 900 km/h for 30 min,   (*b*) 840 km/h for 15 min,
   (*c*) 1200 km/h for $2\frac{1}{2}$ h,   (*d*) 930 mile/h for 40 minutes,
   (*e*) 425 km/h for 10 min and 535 km/h for 20 min.
4. A car travels at 50 km/h for 2 hours and then at 80 km/h for 1 hour. What is the total distance and the average speed for the whole journey?

Travel Graphs 104

Travel Graphs

A journey may be described graphically by plotting the distance travelled against the time. When the speed is constant the distance travelled is proportional to the time taken and the graph is a straight line. The **gradient** of the line is equal to the **speed**.

Example The graph in Figure 1 illustrates a 12 km walk made by a party of girls, starting at 8.00 am, and ending at 1.00 pm.
(a) How far did they walk between: (i) 0800 and 0900, (ii) 0930 and 1130?
(b) How long were each of their rest periods?
(c) For what fraction of the total journey time were they on the move?
(d) What was their average speed: (i) in the first hour, (ii) between 9.30 and 10.30, (iii) for the whole journey including resting time, (iv) for the whole journey ignoring the time spent resting?

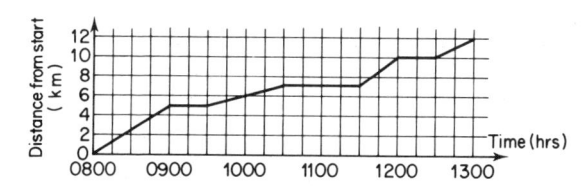

Fig. 1 Travel graph of walk

(a) (i) Between 0800 and 0900 the girls walked **5 km**.
 (ii) Between 0930 and 1130 they walked $(7-5)$ km $= $ **2 km**.
(b) When the graph is flat the girls were not changing their distance from the start and so these are their rest periods:
 0900 to 0930—**half an hour**, 1030 to 1130—**one hour**,
 1200 to 1230—**half an hour**.
(c) Their rest periods totalled $(\frac{1}{2} + 1 + \frac{1}{2}) = 2$ h.
 Total journey time was from 0800 to 1300, 5 hours, so they were on the move for 3 hours out of 5 hours, $\frac{3}{5}$ **of the time**.
(d) (i) During the first hour the girls travelled 5 km, a speed of **5 km/h**.
 (ii) Between 9.30 and 10.30 the girls travelled 2 km, a speed of **2 km/h**.
 (iii) Over the whole journey, the girls travelled 12 km in 5 hours, an average speed of $(12 \text{ km})/(5 \text{ h}) = $ **2.4 km/h**.
 (iv) If the resting time is ignored, the girls travelled 12 km in 3 hours, an average speed of **4 km/h**.

Example Alice's home is 600 m from Bob's. She sets off from home, walking towards Bob's, at 100 m per minute. One minute after she starts, Bob leaves his home and walks at 120 m per minute to meet Alice. Illustrate these journeys on the same travel graph.
(a) How long after Alice left home do they meet?
(b) Are they then nearer to Alice's home or to Bob's and by how much?

105 Travel Graphs

At a speed of 100 m per minute Alice will take 6 minutes to walk the 600 m to Bob's house, so we plan and draw the axes as follows.

Horizontal axis	Time since Alice starts	0–6 min	Scale 2 cm = 1 min.
Vertical axis	Distance from Alice's home	0–600 m	Scale 2 cm = 100 m.

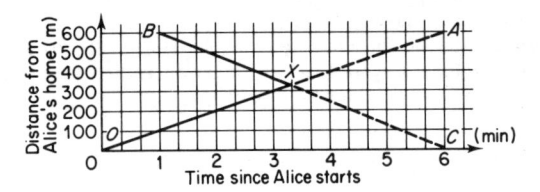

Fig. 2 Travel graph for Alice's and Bob's walk.

The graph for Alice's journey is the line OA, starting from the origin, which would reach Bob's home, distance 600 m from Alice's, after 6 minutes. The graph for Bob's journey, is the line BC, starting from Bob's home, distance 600 m from Alice's, 1 minute after Alice starts, at the point B. Had Bob completed the journey he would have reached Alice's home 5 minutes later at point C, since he walks at 120 m per minute, 600 m in 5 minutes. In fact they meet at the point marked X, at a time **3.3 minutes** after Alice starts. The position of their meeting place is **nearer to Bob's home**, at a distance 330 m from Alice's and 270 m from Bob's, **60 m nearer to Bob's**.

Hints on using Travel Graphs
Draw the time axis across the page and 'distance from some fixed point' axis up the page. Label and scale these axes, and title the whole graph. Choose the scale for time and distance so that: (a) the graph is as large as the paper allows, for greatest accuracy. (b) the 'in between' readings are easy to plot and read off, especially on the time scale when hours and minutes are involved.

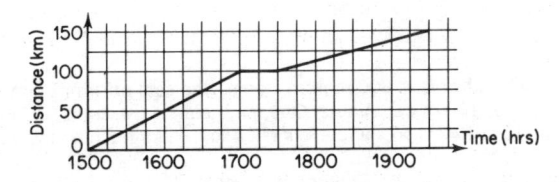

Fig. 3 Car travel graph

SAMPLE QUESTIONS

5 Figure 3 shows a car journey.
  (a) How far had the car travelled by 5.00 pm?
  (b) For how long did the driver stop during his journey?
  (c) What was the average speed of the car during the first hour?
  (d) What was the average speed for the whole journey, including stops?
6 Draw a travel graph for a man who starts with a 25 mile hike, walking at 5 mph and resting for 15 minutes after each 5 miles. He then stops for 2 hours for a meal and returns by train in one hour.
7 A lorry leaves the depot and travels at 50 mile/h for 2 hours. The driver stops at a cafe for half an hour and then continues travelling for one hour at an average speed of 30 mile/h. Draw a travel graph of his journey and state how far he travelled.
8 Bob and Carol drive to visit their sister Anne for lunch, 260 km away. They leave at 9 am and return by 7 pm. Figure 4 shows their journey.

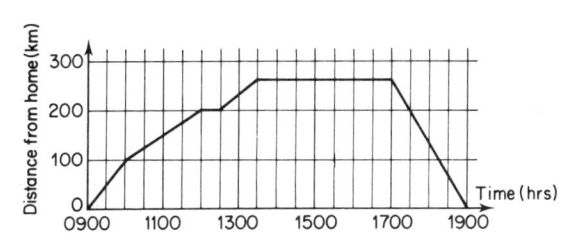

Fig. 4 Bob and Carol's journey

  (a) How far are they from home at: (i) 1100,   (ii) 1230?
  (b) For how long did they stop during their outward journey?
  (c) At what time did they: (i) arrive at Anne's, (ii) leave for home?
  (d) How long did it take them to drive home again?
  (e) What was their average speed for the whole trip: (i) counting the time they stopped on the way, but not the time spent at Anne's, (ii) just counting the time they were moving?
9 Show on a travel graph the journey of Bill to school 4 km away. He cycles steadily at 20 kph. His sister Sheila leaves at the same time, walks the first kilometre in 8 minutes and then gets a lift to school taking only another 3 minutes. Show that Sheila gets to school first. Find how far it was from school when Sheila passed Bill and how long she had to wait at school until Bill arrived.
*10 Fred and Mary swim lengths in a swimming pool 50 m long. They each swim steadily, Fred taking 1 minute and Mary 48 seconds to swim one length. Mary gives Fred a 1 minute start.   (a) When do they first cross?   (b) How many lengths have they done when Mary first catches up with Fred?   (c) How many times did they cross before then?

EQUATIONS AND GRAPHS
6.6 Plotting Functions

6

Calculating tables of values

When evaluating a function for a series of values of x, it is helpful to tabulate **intermediate terms** as shown in the next example.

Example Evaluate the function f, where $f(x) = x^2 - 3x - 2$ for integer values of x in the range -2 to 4, and draw the graph of f.

x	x^2	$-3x$	-2	$x^2 - 3x - 2$
-2	4	6	-2	8
-1	1	3	-2	2
0	0	0	-2	-2
1	1	-3	-2	-4
2	4	-6	-2	-4
3	9	-9	-2	-2
4	16	-12	-2	2

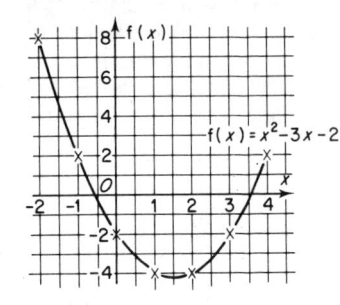

Fig. 1 Table of values and graph of $f(x) = x^2 - 3x - 2$

Set out the table with columns for $x, x^2, -3x, -2$, and the complete function $f(x) = x^2 - 3x - 2$. Include the negative sign in $-3x$ and -2, so that the last column is the sum of the middle three columns.

Work down each column in turn, for instance, to find the entries in the $-3x$ column, multiply each entry in the first column by -3.

Use the patterns in the table to check for errors.

Draw the axes, Ox from -2 to 4, and Oy from -4 to 8, taking the greatest and smallest values from the $f(x)$ column.

Plot the points and **join with a smooth curve,** labelled $f(x) = x^2 - 3x - 2$.

*Example Complete the table of values for the equation $y = 18/(x+1)$ and draw the graph for $x = 1$ to $x = 5$. Use your graph to find an approximate solution of the equation $18/(x+1) = 5.5$

x	1	1.5	2	2.5	3	3.5	4	4.5	5
y	9	7.2							

To help the calculation an extra row $x + 1$ has been inserted in the table. The operations have also been written on the right.

x	1	1.5	2	2.5	3	3.5	4	4.5	5
$x+1$	2	2.5	3	3.5	4	4.5	5	5.5	6
y	9	7.2	6	5.1	4.5	4.0	3.6	3.3	3

$\downarrow +1$

$\downarrow$ divide into 18

Plotting Functions 108

The points are plotted and joined by a smooth curve, as shown in Fig 2. Mark on the graph the line $y = 5.5$ and this crosses the graph of $y = 18/(x+1)$ at $x = 2.3$, (see the two dotted lines), the approximate solution to the equation.

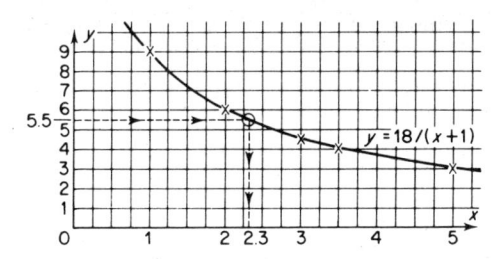

Fig. 2 The graph of $y = 18/(x+1)$ and solution of $18/(x+1) = 5.5$

SAMPLE QUESTIONS

1. The depth of water in a storage tank is 60 cm when full. The volume, V litres, of water in the tank when the depth is d cm is given by $V = 75d/20$. (a) Calculate the value of V when $d = 0, 20, 40, 60$. (b) State the volume of water in the tank when it is full. (c) Plot the points using a scale 2 cm to 10 cm for d on the horizontal axis, and a scale of 1 cm to 20 litres for V on the vertical axis. Join the points and use the graph to find: (i) the volume of water in the tank when the depth is 17 cm, (ii) the depth when there is 120 litres in the tank.

2. Solve the following equations by drawing a graph:
   (a) $15 - 7x = 3$ (b) $x^2 - 6x + 3 = 7$ *(c) $12/(4x - 3) = 5$

3. When rewinding a tape the recorder reading is noted every 5 seconds.

Recorder reading	109	88	61	19
Time after start	0	5	10	15

   Using scales 1 cm to 10 on the reading axis, and 1 cm to 1 second on the time axis, plot these points, join, and extend to meet the time axis. From your graph find the approximate value of: (a) the total time taken for the rewind, (b) the reading on the recorder after 12 seconds.

4. A shop offers H.P. terms repaying over a period of months up to 2 years. For the purchase of £100 of goods the monthly repayments are shown opposite. Plot a graph and use it to estimate: (a) the

Period (months)	6	12	18	24
Repayment (£)	17.80	9.50	6.70	5.30

   monthly repayment over 8 months, (b) the interest paid for repaying over 21 months.

*5. The charge, £C, for a time, t hours, spent repairing washing machines is given in the table opposite. Find a

t	0	0.5	1	1.5	2	2.5
C	15	20	25	30	35	40

   function for C in terms of t, and the hourly charge for repair.

EQUATIONS AND GRAPHS
6.7 Interpreting Graphs   6

A graph is often used to display how something **changes** as time goes on. Typical variables are height, weight, distance, temperature, or speed.

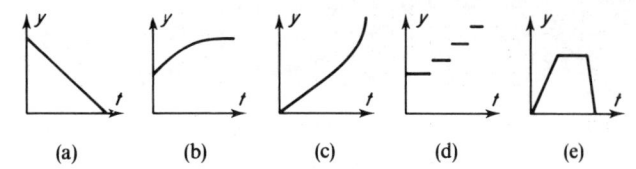

Fig. 1 The variation of y with time t.

Example Describe how y is varying with time, t, in the graphs drawn in Figure 1. In each case, match the graph to one of the following situations: (i) the height of a boy between ages 10 and 20, (ii) the depth of water in a bath as it empties, (iii) the speed of a car moving between traffic lights, (iv) the annual salary of an employee over several years, (v) the depth of fluid in a bottle with a narrow neck as it is being filled steadily.

(a) y decreases steadily with time until it is zero, and this fits the emptying bath (**ii**).
(b) y increases but the rate of increase slows until it is very small, and this fits the growing boy's height (**i**).
(c) y increases steadily for some time and then begins to increase faster towards the end. This matches the filling of a bottle where the narrow neck makes the level rise sharply towards the end (**v**).
(d) y is a **step** function, constant for a time, then jumping to the new level, which fits the employee's annual salary rises (**iv**).
(e) y increases steadily to a maximum, remains constant for some time and then decreases to zero, matching the car's speed between lights (**iii**).

*Gradient
The steepness of a graph, up or down, indicates the rate of change of y with respect to x, and this is known as the **gradient** of the function x to y.

$$\text{gradient} = \frac{\text{increase in } y}{\text{increase in } x}$$

If y is increasing as x increases then the gradient is **positive**, as in Example (ii) and (iii), whilst if y is decreasing as x increases then the gradient is **negative**, as in Example (i) above.

Interpreting Graphs 110

*Example The first graph in Figure 2 shows the travel graph of a car journey. Draw the corresponding speed time graph for the journey.

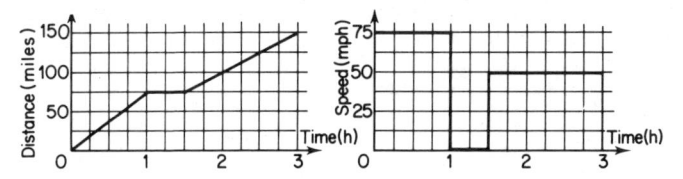

Fig. 2 Distance–time and speed–time graphs for a car journey.

The car travels 75 miles at a steady rate in the first hour, which means a constant speed of 75 mph. For the next half hour the car is at rest, travelling with zero speed. For the next 90 minutes the car steadily covers another 75 miles, which means a speed of $(75/1\frac{1}{2})$ mph = 50 mph. The resulting speed–time graph is also drawn in Figure 2.

When the graph is curved, the gradient is continually changing but it can be **estimated** by measuring the **gradient of the tangent to the curve**.

*Example Figure 3(a) shows the distance–time graph for a model car projected along a level floor. By drawing tangents estimate the speed at 2 second intervals and draw the speed–time graph for the motion.

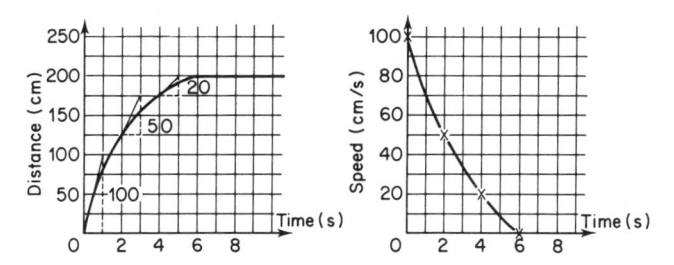

Fig. 3 Estimating speeds by drawing tangents on a travel graph.

Draw tangents to the curve at $t = 0$, 2, 4 and 6.

The gradient of each of the corresponding tangents is measured in cm/s as 100, 50, 20 and zero. The points are plotted on the speed–time axes and joined with a smooth curve.

*Speed–time graphs

The **gradient of a speed–time graph** represents the rate of change of speed with respect to time, known as the **acceleration**.

111 Interpreting Graphs

*__Example__ Figure 4 shows the speed–time graph for a rocket during the first 20 seconds of flight. Find the acceleration of the rocket during, (a) the first 12 seconds, (b) the next 8 seconds of flight.

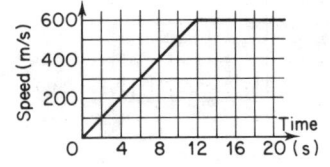

Fig. 4 Speed–time graph of rocket.

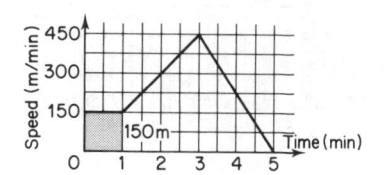

Fig. 5 Speed–time graph of tram.

(a) The rocket increases speed steadily from rest to 600 metres per second over the first 12 seconds. This represents an **acceleration** of (600/12) (metres per second) per second, or **50 ms^{-2}**. The gradient of the first 12 seconds of the speed–time graph is 50, matching this acceleration.
(b) For the next 8 seconds the rocket holds this speed which means the rate of change of speed, the acceleration, is **zero**.

*Area under a speed–time graph

The **area under a speed–time graph** represents the **distance travelled**. The best way to understand this is to start by identifying the distance represented by **one** square under the graph, as shown in the next example.

*__Example__ Figure 5 shows the speed–time graph of a electric tram. Calculate the distance travelled during, (a) the first minute, (b) the second and third minutes, (c) the whole journey.

(a) The tram travels at a steady speed of 150 metres per minute for the first minute, making a distance of **150 m**, so each large square under this graph represents 150 m.
(b) During the second and third minutes the area under the speed–time graph is 4 large squares, representing 4×150 m = **600 m**.
(c) During the last two minutes the area under the graph is 3 large squares, 450 m, so the total distance travelled is $(150 + 600 + 450)$ m = **1200 m**.

SAMPLE QUESTIONS
1 For each of the distance time graphs in Figure 6 sketch the speed–time graph and identify which of the following situations best fit. (a) a stone falling from a cliff, (b) a man running to a bus stop, waiting, and then catching a bus, (c) a billiard ball rolling along the table.

Interpreting Graphs 112

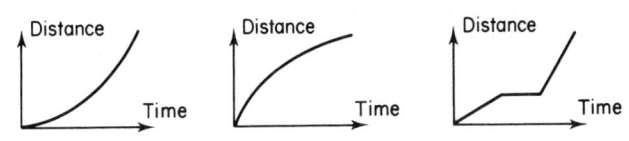

Fig. 6 Distance time graphs.

2 The graph *ABCDEFGHI* in Figure 7 shows the water level whilst Jenny takes a bath. Explain the significance of each segment of the graph.

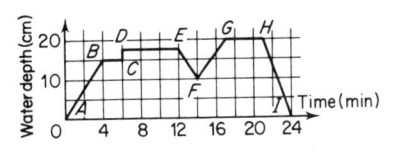

Fig. 7 Jenny's bath.

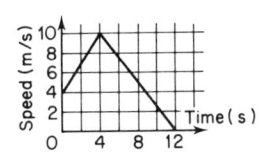

Fig. 8 Speed–time graph.

*3 Figure 8 shows the speed–time graph of a particle. Find: (*a*) the speed of the particle after 2 seconds, (*b*) the acceleration of the particle in the last 8 seconds of the motion, (*c*) the total distance travelled by the particle.

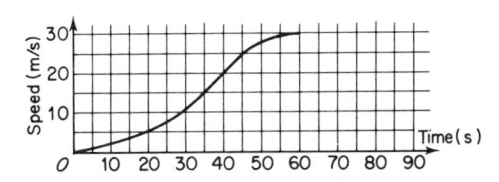

Fig. 9 The speed–time graph of a car.

*4 The speed of a car is observed at regular intervals of time and the speed–time graph drawn in Figure 9 has been drawn from these observations. Use the graph to estimate the speed and acceleration of the car when $t = 20, 40, 60$ seconds after starting. Draw an acceleration time graph for the first 60 seconds of the car's motion. Use the area under the speed–time graph to estimate the total distance travelled by the car. When $t = 60$ the driver applies the brakes to produce a constant retardation of 2 ms^{-2}. Extend the speed–time graph to show this deceleration and state the value of t when the car comes to rest. How far does the car travel during this deceleration?

EQUATIONS AND GRAPHS
6.8 *Composite Functions 6

The function f: $x \to (x+1)^2$ is the **combination of two operations**.
'Add 1' is written as h: $x \to x+1$, and 'square' is written g: $x \to x^2$.
Then f is called the **composite** of the two functions g and h and f = gh, written 'backwards' since gh means operate first by h and then by g.
$f(x) = gh(x) = g(h(x)) = g(x+1) = (x+1)^2$

$$\begin{array}{c} \longrightarrow \quad f \quad \longrightarrow \\ x \xrightarrow{h} x+1 \xrightarrow{g} (x+1)^2 \\ 3 \longrightarrow \quad 4 \longrightarrow 16 \end{array} \qquad \begin{array}{l} h(x) = x+1 \\ g(x) = x^2 \\ f(x) = (x+1)^2 \end{array}$$

*Example Given functions f: $x \to 2x-1$, and g: $x \to x^2+2$ write down: (a) f(3), (b) g(5), (c) gf(3), (d) fg(3). Write the functions fg and gf in the form $x \to \ldots$

(a) $f(3) = 2 \times 3 - 1 = \mathbf{5}$.   (b) $g(5) = 5^2 + 2 = \mathbf{27}$
(c) $gf(3) = g(f(3)) = g(5) = \mathbf{27}$, using the results of parts (a) and (b).
(d) $fg(3) = f(g(3))$. Now $g(3) = 3^2 + 2 = 11$,
so $fg(3) = f(11) = 2 \times 11 - 1 = \mathbf{21}$
$fg(x) = f(x^2 + 2) = 2(x^2 + 2) - 1 = 2x^2 + 3$
$gf(x) = g(2x - 1) = ((2x-1)^2 + 2) = 4x^2 - 4x + 3$
So **fg**: $x \to 2x^2 + 3$, and **gf**: $x \to 4x^2 - 4x + 3$
fg(3) and gf(3) are not equal, so the order of these functions matters.

*Inverse operations and inverse functions

The operation 'add 2' has an **inverse**, or opposite, operation 'take 2'.
The function f: $x \to x+2$ has an **inverse function** $f^{-1}: x \to x-2$.
The function g: $x \to 3x-5$ has an inverse function $g^{-1}: x \to (x+5)/3$.
The order of the inverses of the two operations 'times 3' and 'take 5' has to be reversed, 'add 5' and 'divide by 3', to make the inverse function.

$$g: x \xrightarrow{\times 3} 3x \xrightarrow{-5} 3x - 5 \qquad g^{-1}: x \xrightarrow{+5} x+5 \xrightarrow{\div 3} (x+5)/3$$
$$2 \longrightarrow 6 \longrightarrow 1 \qquad\qquad 1 \longrightarrow 6 \longrightarrow 2$$

However some functions do not have an inverse function, since there is more than one answer when you try to reverse the operation. The function f: $x \to x^2$ has the property $f(1) = 1$ and $f(-1) = 1$. So reversing this function would give two answers for $f^{-1}(1)$, namely 1 and -1. Such a function has **no inverse** and is called a **many-one** function. Functions which have inverse functions are called **one-one** since no two elements are mapped on to the same number.

*Composite Functions

*Example Find the inverse function, where possible: (a) $f: x \to (2x-1)/3$ (b) $g: x \to 8/(2-x)$, $x \neq 2$, (c) $h: x \to \sin x°$

(a) Break up the function f into separate operations, reverse each operation and rebuild f^{-1} in the opposite order.

$$f: x \xrightarrow{\times 2} 2x \xrightarrow{-1} 2x-1 \xrightarrow{\div 3} (2x-1)/3$$

$$f^{-1}: x \xrightarrow{\times 3} 3x \xrightarrow{+1} 3x+1 \xrightarrow{\div 2} (3x+1)/2$$

(b) For the function g the operations are more complicated, using 'change sign', written $+/-$, and 'reciprocate', written $1/x$.

$$g: x \xrightarrow{+/-} -x \xrightarrow{+2} 2-x \xrightarrow{1/x} 1/(2-x) \xrightarrow{\times 8} 8/(2-x)$$

$$g^{-1}: x \xrightarrow{\div 8} x/8 \xrightarrow{1/x} 8/x \xrightarrow{-2} (8/x)-2 \xrightarrow{+/-} -(8/x)+2, \quad x \neq 0.$$

(c) Sin 0° = 0 and sin 180° = 0, so h is a many-one function and therefore has **no inverse**.

SAMPLE QUESTIONS

*1 Given functions $f: x \to 4x+3$, and $g: x \to 1-x^2$, find: (a) f(1), (b) g(1), (c) fg(1), (d) gf(1). Write the two functions fg and gf in the form $fg(x) = \ldots$.

*2 Find the inverse operations to: (a) add 9, (b) take 7, (c) divide by 4, (d) multiply by 2, (e) square root, (f) change sign.

*3 Write the composite function described by the following sequence of operations, in the form $x \to \ldots$: (a) add 2 and then multiply by 7, (b) square and then divide by 3, (c) take 3, change sign, multiply by 9.

*4 Find the inverse function, where possible: (a) $f: x \to 5x+2$ (b) $f: x \to 3-2/x, x \neq 0$ (c) $f: x \to x^2-1$.

*5 The functions f, g and h are defined as follows; $f: x \to x-3$, $g: x \to x/(2x+1)$, $x \neq -\frac{1}{2}$, $h: x \to x^2+2x$.
 (a) Which of the 3 functions is a many-one function?
 (b) Find h(−3) and gf(2),
 (c) Write down, in the form $x \to \ldots$, the composite functions fh and hf.

*6 Calculate a table of values for the function $f: x \to 3x+1$, with values of x in the range −1 to 4. Draw axes, plot and join up to show the function f. Find the inverse function f^{-1}. On the same axes draw the function f^{-1}. Describe a transformation of the plane which will map the graph of f onto the graph of f^{-1}.

EQUATIONS AND GRAPHS
6.9 Linear Functions

Gradient and data zero for $y = mx + c$

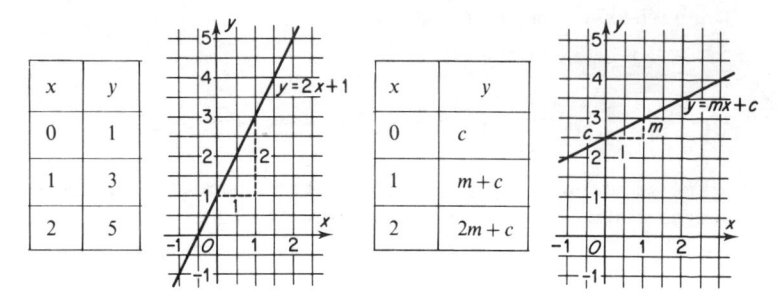

x	y
0	1
1	3
2	5

x	y
0	c
1	$m+c$
2	$2m+c$

Fig. 1 The table of values and the graph of a linear function.

Figure 1 shows the graph of two **linear** (straight line) functions. $y = 2x + 1$, a particular linear function, meets the y-axis at the point (0, 1) and has gradient 2 (see page 116), '1 across 2 up'.

$y = mx + c$, the **general form** for a **linear function**, meets the y-axis at the point (0, c) and has gradient m, '1 across m up'.

c, the **data zero**, is the value of y at the intercept with the y-axis, so may be found by substituting $x = 0$, and finding the value for y.

m, the **gradient**, is the increase in y for a unit increase in x. The height of the triangle with unit base drawn in Figure 1 gives this gradient, positive if the graph is increasing and negative if decreasing.

Example Find the data zero, gradient, and draw the graph of: (a) $y = \frac{1}{2}x - 2$ (b) $3x + 2y = 6$.

(a) $y = \frac{1}{2}x - 2$ is already in the general form $y = mx + c$, with $m = \frac{1}{2}$ and $c = -2$, so the line has gradient $\frac{1}{2}$, and data zero -2, and passes through the point $(0, -2)$ (Figure 2).

(b) $3x + 2y = 6$ can be reorganised into the general form by subtracting $3x$ and dividing both sides by 2.

$$\begin{array}{r|l} & 3x + 2y = 6 \\ -3x\,| & 2y = -3x + 6 \\ \div 2\,\,| & y = -\frac{3}{2}x + 3 \end{array}$$

The gradient is $-\frac{3}{2}$ and the data zero is 3, giving a graph passing through (0, 3), sloping so that '1 across means $\frac{3}{2}$ down' (Figure 2).

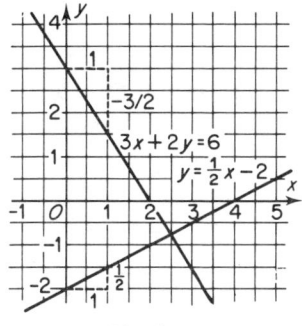

Linear Functions 116

Fig. 2   Fig. 3

Example Find the gradient and data zero, and hence the equation, of the line: (a) passing through the points (1, 3) and (3, −1), (b) passing through (−1, −2) and parallel to the line $y = 3x − 1$.

(a) Figure 3 shows a rough sketch of the line, with a triangle marked on the line between (1, 3) and (3, −1). The slope of the line is '2 across and 4 down' giving a gradient of **−2**, and it is clear that the data zero is **5**, by moving '1 back and 2 up' from (1, 3). The equation of the line is therefore $y = -2x + 5$ or $y + 2x = 5$.

(b) The gradient of $y = 3x − 1$ is 3, so any parallel line will also have gradient 3. From (−1, −2) move '1 across 3 up' to reach the y-axis at (0, 1), so that data zero is **1**. Then the equation of the line is $y = 3x + 1$.

SAMPLE QUESTIONS

1 Draw the graph of the following linear functions:
   (a) $y = x - 3$ (b) $2y - x = 5$ (c) $2x + 3y = 12$ (d) $5 - 3x = y$
   (e) $7 - y = 4x$
2 Find the gradient and data zero of the lines with equations given in question 1.
3 Find the gradient of the line passing through the two points: (a) (2, 0), (3, 4) (b) (3, 2), (5, −4) (c) (3, 1), (−2, 3) (d) (−1, −2), (4, 5)
*4 Find the equation of the line: (a) passing through (4, 1) with gradient 2 (b) passing through the (2, 3) and parallel to $y + x = 0$, (c) passing through (−2, 1) and (3, −4), (d) with gradient $-\frac{1}{2}$ and passing through the origin.
*5 The graph of the line with equation $5x + 12y = 60$ cuts the x-axis at X and the y-axis at Y. Find the coordinates of the points X and Y and calculate the gradient of the line. Find, by Pythagoras' result, the length of XY.
*6 $ABCD$ is a square with vertices $A(0, 0)$, $B(4, 1)$, $C(3, 5)$ and $D(-1, 4)$. Find the equations of the four lines which form the edges of the square.

EQUATIONS AND GRAPHS
6.10 Linear Equations   6

Inverse operations

The equation $x+2 = 7$ may be interpreted as **add 2 to x to make 7**. The inverse operation of **add 2** is **take 2** so **take 2 from 7 to get x**.

This may be written: $x \xrightarrow{+2} x+2$ or $x+2 = 7$

$5 \xleftarrow[-2]{} 7$ $\underline{-2|}\ x = 5$

The first solution uses the **function** and **inverse function** notation, which may help in deciding which operation to use first in complicated examples. The second solution lays the work out in a more traditional form but still shows that the **inverse** operation **take 2** must be applied **to both sides** of the equation in order to find x.

Example Solve the equations: (a) $5x = 30$ (b) $3x - 2 = 19$
(c) $4(x+1) = 12$.

(a) The inverse of **times by 5** is **divide by 5**, so: $5x = 30$
$\underline{\div 5|}\ x = 6$.

(b) The function $x \to 3x - 2$ is formed from the operations **times 3** and **take 2**, so the inverse operations must come in the order **add 2** and **divide by 3**, as follows:

$3x - 2 = 19$
$\underline{+2|}\ \ 3x = 21$
$\underline{\div 3|}\ \ \ x = 7$ Check: $3 \times 7 - 2 = 19 \checkmark$

or in function form:

$x \xrightarrow{\times 3} 3x \xrightarrow{-2} 3x - 2$

$7 \xleftarrow[\div 3]{} 21 \xleftarrow[+2]{} 19$

(c) The function $x \to 4(x+1)$ is formed by **add 1** and **times 4**, so the inverse operations needed are **divide by 4** and **take 1**:

$4(x+1) = 12$
$\underline{\div 4|}\ \ x + 1 = 3$
$\underline{-1|}\ \ \ \ x = 2$ Check: $4(2+1) = 4 \times 3 = 12 \checkmark$

When x occurs on both sides of a linear equation then the function method cannot be used directly, but the rule **do the same to both sides** can still be used to solve the equation. Remember to **check** the answer.

*__Example__ Solve the equations: (a) $4x + 5 = 12 - 3x$ (b) $\frac{1}{4}(2x - 3) = \frac{3}{2}$

Linear Equations 118

In (a) the first inverse operation used is **add 3x**, and then the solution follows as before.

(a) $\quad 4x+5 = 12-3x$ $\qquad$ (b) $\quad \frac{1}{4}(2x-3) = \frac{3}{2}$

$\underline{+3x\,|}\quad 7x+5 = 12 \qquad\qquad \underline{\times 4\,|}\quad 2x-3 = 6$

$\underline{-5\,|}\quad\;\; 7x = 7 \qquad\qquad\quad \underline{+3\,|}\quad\;\; 2x = 9$

$\underline{\div 7\,|}\quad\;\; x = 1 \qquad\qquad\quad \underline{\div 2\,|}\quad\;\; x = \frac{9}{2}.$

Check: $4 \times 1 + 5 = 12 - 3 \times 1,$ $\qquad$ Check: $\frac{1}{4}(2 \times \frac{9}{2} - 3)$
$\qquad\qquad\quad 9 = 9\checkmark \qquad\qquad\qquad\qquad\qquad = \frac{1}{4}(\frac{12}{2}) = \frac{3}{2}\checkmark$

In (b) the first operation is **times by 4** which has the effect of removing all the fractions from the equation.

*Solution of linear inequalities

The same method may be used to solve linear inequalities, except that operations which involve **multiplication** or **division by a negative number** reverse the inequality sign.

*__Example__ Solve the inequality: (a) $3x-2 < 10$ (b) $2(3-x) > 8$

(a) $\quad 3x - 2 < 10$ $\qquad\qquad$ (b) $\quad 2(3-x) > 8$

$\underline{+2\,|}\quad 3x < 12 \qquad\qquad\quad \underline{\div 2\,|}\;\; 3 - x > 4$

$\underline{\div 3\,|}\quad\;\; x < 4 \qquad\qquad\quad\;\; \underline{-3\,|}\;\; -x > 1$

$\qquad\qquad\qquad\qquad\qquad\quad \underline{\times(-1)\,|}\quad x < -1$

SAMPLE QUESTIONS

1. Solve the equations: (a) $x+2 = 10$ (b) $3x+15 = 0$
   (c) $2x+1 = 9$ (d) $5-4x = 21$ (e) $\frac{1}{3}(x+1) = 4$
   (f) $\frac{1}{5}(2x+1) = 3$ (g) $\frac{1}{2}(12-5x) = 11$

*2 Solve the equations: (a) $5(x-1) = 15$ (b) $2(3-5x) = -24$
   (c) $2x+3 = 3x-2$ (d) $x+1 = 3(x-3)$
   (e) $2(x-3) = 3(8-x)$ (f) $(x+2)/3 = x-6$
   (g) $(x-4)/3 = (x+1)/4 + 1.$

*3 Solve the inequalities: (a) $x+4 < 9$ (b) $2(x-3) < 12$
   (c) $x-5 < 3x+1$ (d) $x+4 < 2(x-2)$ (e) $13-5x > 3$
   (f) $3(4-x) > 2(3x+9)$ (g) $\frac{1}{4}(2x+5) < x/3$
   (h) $2(3x-1) < 5x.$

*4 Find the largest integer x such that $5x-7 < 2x+3.$

*5 x and y are related by the equation $3x+4y = 5.$ (a) Given that $x \geqslant 10$, what is the greatest value of y? (b) Find an inequality for x if $y < 1.$

6. James chooses a number, doubles it and then adds 7. The answer is then 111. Find the number James chose.

EQUATIONS AND GRAPHS
6.11 Problem Solving with Algebra

Rearranging formulae

Algebra is often used to express a problem in symbols. These symbols may form an **equation**, or a **formula**, expressing one quantity in terms of another. The equation is then **solved**, or the formula **rearranged**, and the result **interpreted** to provide a solution to the original problem.

Example "Think of a number, add 5, multiply by 3 and take 10. If the answer is 17, what was the number?"

Start by assuming that x is the number. Draw a flow chart to find the answer 17, and then reverse each step to find the original number x.

$$x \xrightarrow{+5} x+5 \xrightarrow{\times 3} 3(x+5) \xrightarrow{-10} 3(x+5)-10$$

$$4 \xleftarrow[-5]{} 9 \xleftarrow[\div 3]{} 27 \xleftarrow[+10]{} 17$$

So the original number was **4**. This process is similar to finding the solution to the equation $3(x+5) - 10 = 17$.

*Example John is going on holiday for 4 days and discovers that the cost of hiring a car is £D per day plus P pence for each mile driven. (a) Write down an expression for the total cost £C in terms of P, D and M, if he travels M miles altogether; (b) Given that the daily charge is £5.50 and the mileage rate is 1.5p per mile, use this formula to find the total cost if he travels: (i) 750 miles, (ii) 1250 miles; (c) Rearrange your formula to express M in terms of C, P, and D; (d) Find how many miles he travelled during his holiday if the total charge was £45.70, and the rates remained the same as in (b).

(a) The daily hire rate is £D, so for 4 days this will cost £$4D$. The mileage rate is P p for 1 mile, so M miles will cost PM p, which is £$PM/100$. The total cost is £C, where $C = £(4D + PM/100)$.

(b) $D = 5.50$, $P = 1.5$, so if $M = 750$, $C = ((4 \times 5.50) + (1.5 \times 750)/100)$ so the cost is £**33.25**.
If $M = 1250$, $C = ((4 \times 5.5) + (1.5 \times 1250)/100)$, so the cost is £**40.75**.

(c) A **formula** may be **rearranged** in the same way as **linear equations are solved**. Consider how the function $M \to 4D + PM/100$ is built up, then reverse each step in turn, so making M the **subject of the formula**.

$$M \xrightarrow{\times P} PM \xrightarrow{\div 100y} PM/100 \xrightarrow{+4D} 4D + PM/100$$

$$100(C-4D)/P \xleftarrow[\div P]{} 100(C-4D) \xleftarrow[\times 100]{} C - 4D \xleftarrow[-4D]{} C$$

So $M = 100(C - 4D)/P$

(d) $C = 45.7$, $P = 1.5$, $D = 5.5$, so substituting:

$M = 100(C - 4D)/P = 100(45.7 - 4 \times 5.5)/1.5 = 1580$.

So he drove **1580 miles** on his holiday.

SAMPLE QUESTIONS

1. After shopping, Angela has x pence left. Betty has 7p less than Angela. Express the amount of money Betty has in terms of x. If Betty really has 25p, how much has Angela?
2. A class has 27 pupils of whom b are boys. Write down an expression for the number of girls in the class. In fact there are twice as many girls as boys in the class. Find how many boys there are in the class.
3. Rearrange $y = 3x - 7$ to make x the subject of the formula.
4. The final speed v of a body is given by the formula $v = u + at$, where u is the initial speed, a the acceleration, and t the time taken. (a) Find v when $u = 20$, $a = -12$ and $t = 3$. (b) Find t when $u = 10$, $v = 30$ and $a = 5$. (c) Find the acceleration of a body which goes from rest to 28 m/s in 4 seconds.
5. n is a whole number. Write down an expression for the next whole number bigger than n. If these two numbers add up to 27 find the value of n.
6. 151 people live in a village and m are men. There are 17 more women than men and 30 more children than women. Write down, in terms of m, the number of women and the number of children. Find how many men, women and children live in the village.
*7. A room measures 3 m by 4 m. A carpet is to be chosen so that there is an uncarpeted border, x m wide all around the carpet. Write down, in terms of x: (a) the length and breadth of the carpet, (b) the area of the carpet, (c) the uncarpeted area.
*8. George is g years old. He is 3 years older than his wife and 10 times as old as his son. Their ages add up to 81 years. Find the ages of all the family.
*9. Mary has a 80 m ball of string. She uses it to mark out the perimeter of a rectangular plot of land which is 6 m longer than it is wide. She finds that she uses 3 times as much string as she has left over. Find the dimensions of her plot.
*10. Henry goes to buy stamps. He buys twice as many second class stamps (at 12p) as first class stamps (at 18p). If he buys x first class stamps how much does he spend: (a) on first class stamps, (b) on second class stamps, (c) in total. Find the greatest number of stamps that he can buy if he has £5 to spend.

EQUATIONS AND GRAPHS
6.12 Simultaneous Equations 6

Graphical solution
An equation in two variables, x and y, can be represented as a graph with coordinate axes Ox, Oy. Two such equations give a pair of graphs and the coordinates, (x, y), of any point of intersection of these graphs will be a solution of both equations **simultaneously**.

Example Solve graphically the simultaneous equations:
$2x + 3y = 9$, $3x - 2y = 7$.

The two equations are linear, giving straight lines which may be drawn by the methods given on page 115. Figure 1 shows a table of values for each equation, the points plotted, joined, and labelled. The point (3, 1) where the two lines meet is a solution of both equations simultaneously, so $x = 3, y = 1$ is the solution. This may be checked by noting that (3, 1) is in both tables of values, or by substitution.

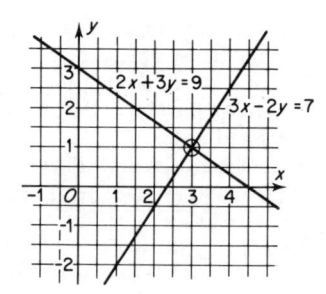

x	y	x	y
0	3	0	$-3\frac{1}{2}$
1	$2\frac{1}{3}$	1	-2
2	$1\frac{2}{3}$	2	$-\frac{1}{2}$
3	1	3	1
4	$\frac{1}{3}$	4	$2\frac{1}{2}$

Fig. 1 Solution of simultaneous equations $2x + 3y = 9$, $3x - 2y = 7$.

*Algebraic solution
Linear simultaneous equations may be solved by **eliminating each variable in turn**.

Simultaneous Equations

Example Solve (a) $2x + 2y = 6$
$\phantom{\text{Solve (a) }}3x + 4y = 8$

(b) $2x + 3y = 9$
$\phantom{\text{(b) }}3x - 2y = 7$

(a) We may eliminate y by doubling the first equation and taking the second.

$$2x + 2y = 6 \quad \ldots \text{①}$$
$$3x + 4y = 8 \quad \ldots \text{②}$$
$$\underline{\text{①} \times 2} \mid 4x + 4y = 12 \quad \ldots \text{③}$$

$\underline{\text{③} - \text{②}} \mid \quad x = 4$
Substituting $x = 4$ in ② gives
$$12 + 4y = 8$$
$\underline{-12} \mid \quad 4y = -4$
$\underline{\div 4} \mid \quad y = -1$
Solution **$x = 4$ and $y = -1$**.
Check:
① $2 \times 4 + 2 \times (-1) = 6 \checkmark$
② $3 \times 4 + 4 \times (-1) = 8 \checkmark$

(b) In this case to eliminate y we must times the first equation by 2, the second by 3 and add.

$$2x + 3y = 9 \quad \ldots \text{①}$$
$$3x - 2y = 7 \quad \ldots \text{②}$$
$\underline{\text{①} \times 2} \mid \quad 4x + 6y = 18 \ldots \text{③}$
$\underline{\text{②} \times 3} \mid \quad 9x - 6y = 21 \ldots \text{④}$
$\underline{\text{③} + \text{④}} \mid \quad 13x = 39$
$\underline{\div 13} \mid \quad x = 3$
Substituting $x = 3$ in ① gives
$$6 + 3y = 9$$
$\underline{-6} \mid \quad 3y = 3$
$\underline{\div 3} \mid \quad y = 1$
Solution **$x = 3$ and $y = 1$**.
Check: ① $2 \times 3 + 3 \times 1 = 9 \checkmark$
② $3 \times 3 - 2 \times 1 = 7 \checkmark$

Clearly it is better to eliminate x first when it is easier to make the coefficients of x the same.

*__Example__ The function f is given by $f: x \to 3x - 2$. Draw the graph of the function f for $-1 \leqslant x \leqslant 5$. Solve by drawing further lines on your graph the equations (a) $3x - 2 = 5$ (b) $3x - 2 = 4 - x$ (c) $3x - 2 = 3x + 2$

Figure 2 shows the table of values and graph of the function f.
(a) $3x - 2 = 5$, may be solved by drawing the line $f(x) = 5$ on the same axes and find the value of x at the point where the two lines meet, which from the graph is **$x = 2.3$ (1DP)**.

x	$3x - 2$
-1	-5
1	1
2	4
3	7
4	10
5	13

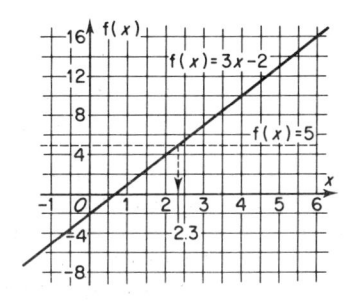

Fig. 2 Function $f(x) = 3x - 2$, and the solution of $3x - 2 = 5$.

Simultaneous Equations

(b) $3x - 2 = 4 - x$ can be solved by drawing the graph of $f(x) = 4 - x$ on the same axes. The table of values and the graph of $f(x) = 4 - x$ are drawn together with $f(x) = 3x - 2$ in Figure 3, and the solution $x = 1.5$ is found from the point where the two graphs meet.

(c) Figure 3 also shows the graph of $f(x) = 3x + 2$, which is parallel to $f(x) = 3x - 2$, but shifted up by 4. Therefore these lines never meet and so there is **no solution** to the equation $3x - 2 = 3x + 2$.

x	$4 - x$
-1	5
0	4
1	3
2	2
3	1
4	0
5	-1

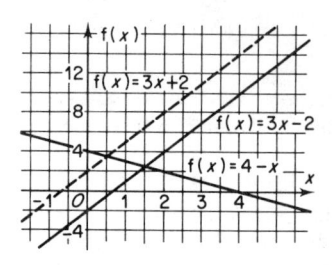

Fig. 3 Solution of $3x - 2 = 4 - x$ and $3x - 2 = 3x + 2$.

*Non-linear simultaneous equations

The next example concerns the simultaneous solution of a quadratic and a linear equation. We use a graphical method here.

*__Example__ Figure 4 is a rough sketch showing the path of a shell fired up a sloping range, with a gradient of $\frac{1}{3}$. Taking axes at the firing point the equation of the path of the shell is given as $y = 1.5x - 0.02x^2$, where x and y are both measured in metres. Draw an accurate diagram showing the path of the shell. Find the coordinates of the point where the shell lands and the range up the slope.

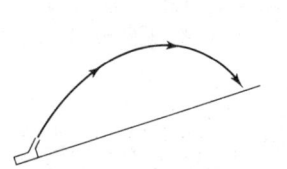

Fig. 4 The shell's path.

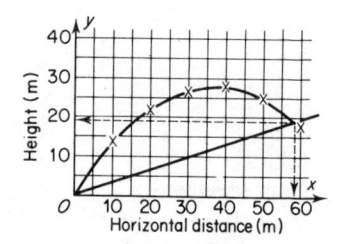

Fig. 5 $y = 1.5x - 0.02x^2$ and $y = x/3$.

The hill is a line passing through the origin, data zero 0, with gradient $\frac{1}{3}$, giving an equation $y = \frac{1}{3}x$, or $y = x/3$.

Calculate a table of values for the shell's path and for the hill.

Scale and label two axes, plot the points and join to give the two graphs given in Figure 5. Reading from the graph the coordinates of the point where the shell lands on the hill is approximately **(58, 19)**.

Simultaneous Equations 124

x	0	10	20	30	40	50	60	
1.5x	0	15	30	45	60	75	90	
$-0.02x^2$	0	-2	-8	-18	-32	-50	-72	
$1.5x - 0.02x^2$	0	13	22	27	28	25	18	... shell's path
x/3	0	3.3	6.7	10	13.3	16.7	20	... the hill

The range up the slope is the distance from (0, 0) to (58, 19) which may be calculated by Pythagoras' result as $\sqrt{(58^2 + 19^2)}$ m = **61 m (2SF)**.

SAMPLE QUESTIONS
1. Solve the following simultaneous equations graphically:
   - (a) $x + 2y = 7$
     $3x - y = 9$
   - (b) $3x - 2y = 6$
     $x + 4y = 8$
   - (c) $2x + 5y = 9$
     $6x + y = -2$
   - (d) $y = 3x - 6$
     $3x + 2y = 8$

*2 Solve the following pair of simultaneous equations algebraically:
   - (a) $x + y = 4$
     $x - y = 2$
   - (b) $x + 2y = 8$
     $2x + 3y = 14$
   - (c) $2x - y = 12$
     $3x + 5y = 5$
   - (d) $3x - y = -1$
     $x + 4y = 17$
   - (e) $3x - 2y = 1$
     $5x - 4y = 3$
   - (f) $5x + 7y = 14$
     $3x - 6y = 39$
   - (g) $2y - 5x = 0$
     $3x + 4y = 26$
   - (h) $4x = 6 - 2y$
     $y = 5x - 4$

*3 Whilst queuing for tickets I noted that the lady in front of me bought 5 adult and 3 children's tickets for £17. I then bought 3 adult and 2 children's tickets for £10.50. Let x be the cost of an adult ticket, y be the cost of a children ticket, and express each of these two statements as an equation in x and y. Solve the two equations to find the price of each ticket.

*4 Draw the graphs of $y = 5 - x$ and $y = x^2 + 3$, on the same axes. Use your graph to find the solutions of the two equations simultaneously.

*5 The functions f and g are given by $f(x) = x^2 - 3x$ and $g(x) = x + 5$. Find the values of x for which $f(x) = g(x)$.

*6 Solve the equations simultaneously by use of a graph:
   - (a) $y = 2x$
     $y = x^2$
   - (b) $y = 4x - 7$
     $y = 2x^2 - 5$
   - (c) $y = 2x + 3$
     $y = x^2 + 2x - 1$
   - (d) $y = 12/(x + 2)$
     $y = x - 1$

*7 The function f is defined by f: $x \to 2 + 8/x$, $(x \neq 0)$.
Evaluate f(1), f(2), f(4) and f($\frac{1}{2}$). Find x when $f(x) = x$.

EQUATIONS AND GRAPHS
6.13 Quadratic Equations
6

Solving by factorisation
Consider a quadratic equation where the quadratic expression can be **factorised**, written as the product of two linear factors. The equation states that the product of these two factors is zero, which occurs only if either of the two factors is zero, giving **two** alternative solutions.

***Example** Solve these equations by factorisation: (a) $(x+1)(x-2) = 0$ (b) $x^2 - 5x + 6 = 0$ (c) $x^2 + x = 12$ (d) $x^2 + 4x + 4 = 0$

(a) $(x+1)(x-2) = 0$ means that either $x+1 = 0$ or $x-2 = 0$, so either $x = -1$ or $x = 2$, giving a solution set $\{-\mathbf{1}, \mathbf{2}\}$.

(b) Factorising the quadratic $x^2 - 5x + 6$ (see page 97) gives $x^2 - 5x + 6 = (x-3)(x-2) = 0$.
Then either $(x-3) = 0$ or $(x-2) = 0$, so **either $x = 3$ or $x = 2$**.

(c) Bring all the terms to the left hand side by taking 12 from both sides. Factorising the quadratic expression $x^2 + x - 12$, gives $x^2 + x - 12 = (x-3)(x+4) = 0$.
Then either $x - 3 = 0$ or $x + 4 = 0$, so **either $x = 3$ or $x = -4$**.

(d) Factorising $x^2 + 4x + 4$ gives two equal factors $(x+2)(x+2)$. Therefore the only solution is $x + 2 = 0$ which means $x = -\mathbf{2}$. This is called a **repeated solution**.

Solving by a graph
When the quadratic expression does not factorise then one method of solving the equation approximately is by drawing a graph and reading off the solutions.

***Example** Solve the equation, (a) $2x^2 - 6x - 5 = 0$ (b) $4x - x^2 = -3$ by using a graph.

(a) Calculate a table of values, selecting the x values in this example so that the quadratic expression changes sign twice to give two solutions.

x	-2	-1	0	1	2	3	4	5
$2x^2$	8	2	0	2	8	18	32	50
$-6x$	12	6	0	-6	-12	-18	-24	-30
$2x^2 - 6x - 5$	15	3	-5	-9	-9	-5	3	15

Plot these points on suitably scaled axes, join up and label the quadratic curve $y = 2x^2 - 6x - 5$ (Figure 1).
Note that $2x^2 - 6x - 5 = 0$ at two points on the x-axis and the approximate values are $x = -\mathbf{0.7}$ **and** $x = \mathbf{3.7}$.

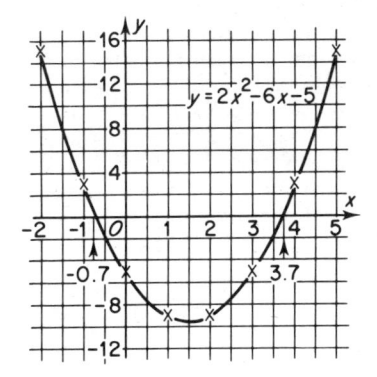

Fig. 1 Solution of the quadratic equation $2x^2 - 6x - 5 = 0$.

(b) We first transfer the -3 term to the other side by adding 3 to both sides, so that the equation becomes $4x - x^2 + 3 = 0$.
Calculate a table of values as before.

x	-2	-1	0	1	2	3	4	5	6
$4x$	-8	-4	0	4	8	12	16	20	24
$-x^2$	-4	-1	0	-1	-4	-9	-16	-25	-36
$4x - x^2 + 3$	-9	-2	3	6	7	6	3	-2	-9

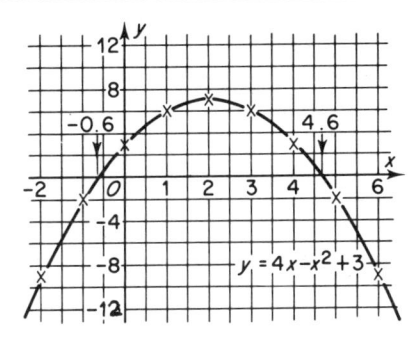

Fig. 2 Solution of the quadratic equation $4x - x^2 = -3$.

Plot these points on the axes as shown in Figure 2, join up and label the curve $y = 4x - x^2 + 3$.
Then the solutions to the equation $4x - x^2 + 3 = 0$ are the x values where the curve cuts the x-axis, which are $x = -\mathbf{0.6}$ **and** $x = \mathbf{4.6}$.

Solving by numerical approximation

A better approximation than is possible from the graph can now be made by **numerical approximation**. We are looking for the value of x when the quadratic changes from positive to negative or back again.

Example Find a 2 decimal place approximation to each of the two solutions of $2x^2 - 6x - 5 = 0$.

We use a calculator to evaluate $q(x)$, where $q(x) = 2x^2 - 6x - 5$.

First Solution		Second Solution	
trial value x $\quad$ q(x)	lies between	trial value x $\quad$ q(x)	lies between
−0.7 $\quad$ 0.18	−0.7 and 0	3.7 $\quad$ 0.18	3.0 and 3.7
−0.6 $\quad$ −0.68	−0.7 and −0.6	3.6 $\quad$ −0.68	3.6 and 3.7
−0.65 $\quad$ −0.255	−0.7 and −0.65	3.65 $\quad$ −0.255	3.65 and 3.7
−0.68 $\quad$ 0.005	−0.68 and −0.65	3.68 $\quad$ 0.005	3.65 and 3.68
−0.67 $\quad$ −0.082	−0.68 and −0.67	3.67 $\quad$ −0.082	3.67 and 3.68

So, to 2 decimal places, the solutions are $x = -\mathbf{0.68}$ and $x = \mathbf{3.68}$.

*Solving by the formula

The general formula for the solution of a quadratic $ax^2 + bx + c = 0$, is:

$$x = \frac{-b \pm \sqrt{b^2 - 4ac}}{2a} \quad \text{so long as} \quad b^2 \geqslant 4ac.$$

If $b^2 < 4ac$ then the quadratic equation has no solutions.

***Example** Solve the following equations by the formula:
(a) $3x^2 - 5x + 2 = 0$ (b) $5x^2 + 3x - 9 = 0$ (c) $2x^2 + 4x + 3 = 0$

(a) Substituting $a = 3, b = -5, c = 2$, into the formula gives,

$$x = \frac{-(-5) \pm \sqrt{(-5)^2 - 4 \times 3 \times 2}}{2 \times 3} = \frac{5 \pm \sqrt{1}}{6} = \mathbf{1 \text{ or } \tfrac{2}{3}}.$$

(b) Substituting $a = 5, b = 3, c = -9$, into the formula gives;

$$x = \frac{-3 \pm \sqrt{3^2 - 4 \times 5 \times (-9)}}{2 \times 5} = \frac{-3 \pm \sqrt{189}}{10}$$

$$= \mathbf{1.075 \text{ or } -1.675 \text{ (3DP)}}$$

(c) Substituting $a = 2$, $b = 4$, $c = 3$, into the formula gives:

$$x = \frac{-4 \pm \sqrt{4^2 - 4 \times 2 \times 3}}{2 \times 2} = \frac{-4 \pm \sqrt{-8}}{4}$$ which cannot be found.

In this case $b^2 < 4ac$, and so the quadratic has **no solution**. This corresponds to a quadratic curve lying totally above the x-axis. So the quadratic expression is always positive and never equal to zero.

SAMPLE QUESTIONS

*1 Solve the equations by factorisation:
   (a) $(x - 3)(x + 1) = 0$ (b) $(2x + 5)(x - 4) = 0$
   (c) $(3x - 1)(5x + 1) = 0$ (d) $x^2 - 9x + 8 = 0$
   (e) $x^2 + 2x - 15 = 0$ (f) $x^2 - 4x - 5 = 0$
   (g) $x^2 - 2x - 63 = 0$ (h) $x^2 + 7x - 44 = 0$
   (i) $x^2 - 49 = 0$ (j) $2x^2 - 3x + 1 = 0$
   (k) $3x^2 + 4x - 15 = 0$ (l) $6x^2 - 7x - 20 = 0$

2 Solve the equations, accurate to 1 DP, by drawing a graph:
   (a) $x^2 - 8x + 8 = 0$ (b) $x^2 + 3x - 15 = 0$ (c) $x^2 - 3x - 5 = 0$

*3 Calculate a table of values of f(x), where f(x) = $2x^2 - 2x - 1$, and draw the graph of the function f for x in the range $-3 \leqslant x \leqslant 4$. Find two approximate solutions to the equation f(x) = 0 from your graph and then use a numerical approximation method to improve the accuracy of your solutions correct to 2 decimal places.

*4 Repeat question 3 for the following functions, choosing a suitable range of values of x:
   (a) f(x) = $x^2 + 3x - 5$ (b) f(x) = $x^2 - 4x + 2$
   (c) f(x) = $2x^2 + 3x - 4$ (d) f(x) = $4x + 1 - x^2$
   (e) f(x) = $3x^2 - 2 - 2x$ (f) f(x) = $7x - x^2 - 4$

*5 Solve $(x - 2)^2 = 9$, by taking the square root of each side of the equation, (remember there should be two solutions).

*6 Solve the following equations using the method of question 5:
   (a) $(x + 3)^2 = 16$ (b) $(2x - 5)^2 = 36$
   (c) $(5 - 2x)^2 = 81$ (d) $(3x - 4)^2 = 10$
   (e) $(2x + 7)^2 = 5$ (f) $(8 - 3x)^2 = 19$

*7 The function f is defined by f: $x \to x^2 - 6$ for all values of x. Find the values of x for which, (a) f(x) = 10,   (b) f(x) = x.

*8 Use the formula method to find the solutions, if any, of the quadratic equations: (a) $x^2 - 7x + 9 = 0$ (b) $x^2 + 4x - 11 = 0$
   (c) $x^2 + 3x + 5 = 0$ (d) $3x^2 - 5x + 8 = 0$ (e) $2x^2 - 7x - 4 = 0$.

EQUATIONS AND GRAPHS
6.14 Linear Inequalities

6

Intervals on the real number line
It is useful to be able to specify a **subset** of the real numbers, such as: 'all the numbers less than -2', or 'all the numbers from 3 to 4'. These regions can be shown by shading the region of the real number line, and described by inequality symbols: $x < -2$, or $3 \leqslant x \leqslant 4$.

Example (a) State the inequalities which represent the shaded regions on the real number line shown in Figure 1(a). (b) Indicate by shading on the real number line the region satisfying: (i) $x \leqslant 2$, (ii) $-1 < x \leqslant 2$ or $3 < x < 4$, (iii) $x \geqslant 0$ and $x \leqslant 3$.

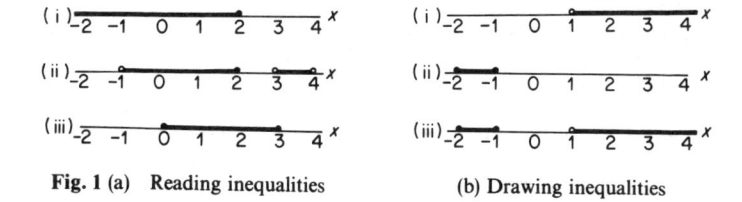

Fig. 1 (a) Reading inequalities   (b) Drawing inequalities

(a)   (i)   The shaded region is all the numbers greater than 1, written $x > 1$.
       (ii)   The region is the numbers from -2 to -1, including the endpoints, so $x \geqslant -2$ **and** $x \leqslant -1$, which can be written as a **double inequality** $-2 \leqslant x \leqslant -1$.
       (iii)   The region is (i) or (ii), so it is written $x > 1$ **or** $-2 \leqslant x \leqslant -1$.

(b)   (i) $x \leqslant 2$, is the set of numbers 2 and below, shaded in Figure 1(b)(i).
       (ii) $-1 < x \leqslant 2$ or $3 < x < 4$ is shown as two regions in Figure 1(b)(ii).
       (iii) $x \geqslant 0$ and $x \leqslant 3$, is the same as $0 \leqslant x \leqslant 3$, shown in Figure 1(b)(iii).

*Inequalities in two variables
Inequalities in two variables x and y can be shown as regions in the Cartesian plane, Oxy. It is common practice to **shade the unwanted region** when considering the region satisfied by several inequalities.

*__Example__   Indicate by shading the unwanted region, the area of the Oxy plane satisfied by the three inequalities, $x \geqslant 2$, $y \geqslant -1$, $x + y \leqslant 5$.

Linear Inequalities 130

The edges of the region are lines found by replacing inequality by equality symbols.
$x \geqslant 2$ is the region on the right of $x = 2$, a vertical line through (2, 0).
$y \geqslant -1$ is the region above the line $y = -1$, a horizontal line through (0, -1).
$x + y \leqslant 5$ is the region on the left and below
$x + y = 5$, a line passing through (5, 0), (0, 5).
The three unwanted regions are shaded in Figure 2 leaving the triangular area clear.

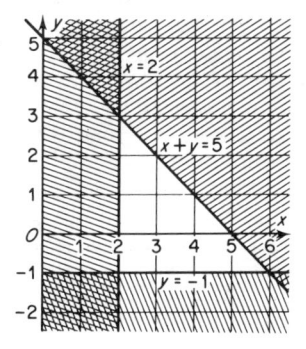

Fig. 2 $x \geqslant 2, y \geqslant -1, x + y \leqslant 5$.

In some applications we may be looking for all **real numbers** satisfying the inequalities. In others we may only be concerned with the **integer** solutions satisfying the inequalities.

*Example On the Cartesian plane, shade those regions where the three inequalities are *NOT* satisfied: $x + 2y \leqslant 10$, $9x + 4y \geqslant 36$, $x \leqslant 2y$. List the integer coordinate pairs (x, y) satisfying all three inequalities.

The boundary lines for these regions will be the lines $x + 2y = 10$, $9x + 4y = 36$, $x = 2y$. Find a table of values for each line and then select two further points, one **satisfying** and one **not satisfying** the given inequality (Figure 3(a)). The points on the lines may now be plotted and joined, and the two further points used to indicate which regions are to be shaded.

Figure 3 shows these lines and regions and the unshaded triangle contains three integer coordinate points, **(3, 3)**, **(4, 2)** and **(4, 3)**.

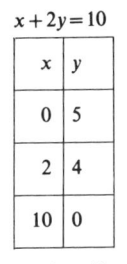

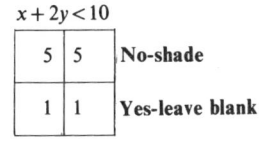

$x + 2y = 10$

x	y
0	5
2	4
10	0

$x + 2y < 10$

5	5	No-shade
1	1	Yes-leave blank

Fig. 3 (a) Table of values

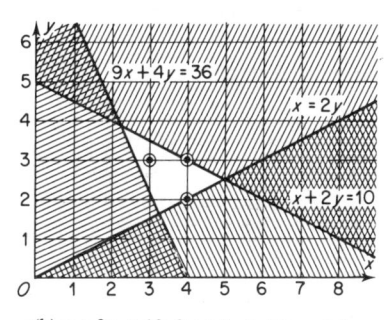

(b) $x + 2y \leqslant 10, 9x + 4y \geqslant 36, x \leqslant 2y$.

*Linear programming

Decisions must often be made to try and get the best results given certain constraints. These constraints may often be expressed as linear inequalities and **linear programming** is the use of graphical methods to find the best solution to such problems, the cheapest cost, the greatest output or the shortest time.

***Example** Oranges cost 9p and peaches cost 12p each. I have £1.20 to spend on a fruit salad and I want to use at least 3 oranges and more peaches than oranges. Assuming that I buy x oranges and y peaches, express these conditions as inequalities and show that in the Oxy plane there are 8 possible decisions that could be made. From each orange I can get 200 g of usable fruit and from each peach 150 g. How many of each must I buy for the largest amount of usable fruit?

The cost of 1 orange is 9p, so x oranges cost $9x$p.
The cost of 1 peach is 12p, so y peaches cost $12y$p.
To keep the total cost below £1.20 I need $9x + 12y \leqslant 120$.
To use at least 3 oranges I need $x \geqslant 3$.
To use more peaches than oranges I need $y > x$.
The three boundary lines are drawn in Figure 4 and the unwanted regions shaded as in the previous example. Note that the $y = x$ boundary is not part of the region and so is shown dotted. There are **8** integer solutions marked with a small ring in the clear triangular region.

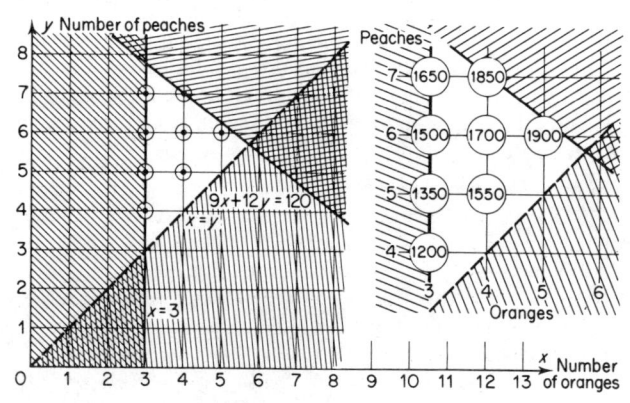

Fig. 4 Choosing oranges and peaches for a fruit salad.

On the right of Figure 4 a larger version of this region is drawn, and within each ring the amount of usable fruit in grams has been calculated. Two sample calculations are shown in the table on page 132.

From the diagram it is clear that the best decision is to buy **6 peaches and 5 oranges** giving 1900 g of usable fruit.

Linear Inequalities 132

Oranges (x)	Peaches (y)	Usable fruit in grams
3	5	$3 \times 200 + 5 \times 150 = 1350$
4	7	$4 \times 200 + 7 \times 150 = 1850$

SAMPLE QUESTIONS

1 Find the inequalities given by the shaded part of the line below.

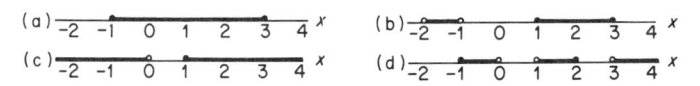

(a), (b), (c), (d) number lines from −2 to 4

2 Indicate by shading on the real number line, the inequalities:
   (a) $2 \leqslant x \leqslant 4$ (b) $-1 \leqslant x \leqslant 1$ or $x > 3$
   (c) $x < 4$ and $x \geqslant -2$ (d) $x < -1$ or $x > 2$

3 State the integer values of x satisfying the inequalities:
   (a) $-1 \leqslant x < 2$ (b) $0.5 < x < 3.5$ (c) $-4.8 < x < -1.8$
   (d) $x > 0$ and $x < 2$

*4 Shade the unwanted regions of the Cartesian plane to show the region where the following inequalities are satisfied. List the integer coordinate solutions of these inequalities.
   (a) $2 \leqslant x \leqslant 4$, $1 \leqslant y < 3$ (b) $x > 0$, $y > 0$, $x + y < 3.5$
   (c) $x \geqslant 2$, $y \leqslant 4$, $x < y + 1$ (d) $x < 5$, $y < 6$, $x + y > 7$
   (e) $x \leqslant y + 1$, $y \leqslant x + 1$, $x + y \geqslant 1$, $x < 2$

*5 I have room in my garden for 7 trees and I want to plant x apple trees costing £15 each, and y pear trees costing £10 each. I must have at least 2 trees of each sort to ensure fertilisation and I have £90 to spend. Write down 4 inequalities in x and y and represent these by shading the unwanted regions in the Cartesian plane. I estimate that in three years time I can expect to get £7 of apples from each apple tree and £5 of pears from each pear tree. Calculate for each of the possible decisions the value of the fruit I expect to get, and indicate the best decision to take to ensure the maximum return.

*6 Jean wants to buy some model horses and pigs for her farm. She has decided to buy at least 3 horses, more pigs than horses and at most 12 animals. Draw a diagram to show the possible purchases she can make. If the horses cost 20p each and the pigs 10p each, find (a) the least amount she can spend,   (b) the largest amount she can spend.

*7 A baker has shelf room for 120 loaves. He bakes x white loaves and y brown, makes a profit of 10p on each white loaf and 5p on a brown loaf, and he knows that he sells at least twice as many brown loaves as white. How many of each should he bake for the greatest profit?

STATISTICS
7.1 Collecting and Sorting Data

In today's world we are inundated with information. For this to be useful it has to be collected and organised in a sensible way. The original information, **raw data**, is often produced by, or **sorted** into a **tally table**. Marks are made, one for each item, grouped in fives for ease of counting, and then added up to give the **frequency** of each result.

Averages: mean, mode, median

An **average** is a single value, a **statistic**, which represents a complete set of data. The **mean** is the average used most often, found by adding up all the values and dividing by their number. Two further averages are the **mode**, the **most common** value, and the **median**, the **middle** value, when the values are placed in order of size.

Example The two sets of numbers $\{a, 7, 3, 9, 2\}$ and $\{4, 8, 3, 5\}$ have the same mean. Find the value of a and the median of each set of numbers.

The sum of the four numbers 4, 8, 3, 5 is 20, so their mean is $20/4 = 5$. Since the mean of the set of 5 numbers $a, 7, 3, 9, 2$ is also 5, their sum must be $5 \times 5 = 25$. Therefore $a + 7 + 3 + 9 + 2 = 25$ and $a = \mathbf{4}$.

Order each set first, then the median of $\{2, 3, 4, 7, 9\}$ is **4**, and the median of $\{3, 4, 5, 8\}$ is **4.5**, the average of the **two** middle numbers.

Example Anita and Mary carried out a survey of the number of people in cars travelling past their school. They produced the table:

People in each car	Tally	Total
1	~~1111~~ ~~1111~~ 111	
2	~~1111~~ ~~1111~~ ~~1111~~ 11	
3	~~1111~~ ~~1111~~	
4	~~1111~~ 1	
5	11	

(a) Complete the 'Total' column. (b) How many cars did they observe?
(c) How many people travelled in these cars?
(d) Write down the modal number of people in a car from this sample.
(e) Calculate the mean number of people in a car in this sample.

(a) Using the tally marks the totals are:
$5+5+3 = \mathbf{13}$ $5+5+5+2 = \mathbf{17}$ $5+5 = \mathbf{10}$ $5+1 = \mathbf{6}$ **2**

(b) The total number of cars is the sum of the 'Totals' column.
$13 + 17 + 10 + 6 + 2 = \mathbf{48}$ cars.

(c) The number of people travelling in these cars is
$(13 \times 1) + (17 \times 2) + (10 \times 3) + (6 \times 4) + (2 \times 5)$
$= 13 + 34 + 30 + 24 + 10 = \mathbf{111}$ people, since, for instance, the 10 cars each with 3 people total $10 \times 3 = 30$ people.

(d) The mode is found from the greatest frequency, 17, which corresponds to 2 people in a car. So the mode is **2**, not the frequency 17.

(e) A total of 111 people travelled in 48 cars, so the mean number of people in a car is 111/48 = 2.3125 = **2.3 (2SF)**, rounding as usual.

SAMPLE QUESTIONS

1. The letters collected from a postbox on one day were classified according to the type of address, by: A(venue), C(lose), R(oad), S(treet), O(ther). Tally the following letters and complete a frequency table.

```
S S R A R A S R O R O S A C R O C C S S R A R S
O S O C R A R O C S A S S A R O C S O R S S O C
A R S C R S S A R S R C O R O S S A O O S R R S
```

Find: (a) the most common type of address, (b) the fraction of all the letters collected which used that type of address.

2. The result of a traffic survey showing the vehicles passing the survey point in one hour is given in the table below.

Vehicle		Number of vehicles
bicycle	̶H̶H̶t̶ ̶H̶H̶t̶ 11	12
motorbike	̶H̶H̶t̶ 111	
car	̶H̶H̶t̶ ̶H̶H̶t̶ ̶H̶H̶t̶ ̶H̶H̶t̶ ̶H̶H̶t̶ 1	
light van	̶H̶H̶t̶ ̶H̶H̶t̶	
heavy van	̶H̶H̶t̶ ̶H̶H̶t̶ ̶H̶H̶t̶	
tanker	1111	
bus	̶H̶H̶t̶ 11	

(a) Fill in the heading for the middle column. (b) Complete the third column, (c) Find how many vehicles passed the point in the hour.

3. Jacob measured the weekly rainfall over 6 weeks. His results, in centimetres, were 6, 10, 9, 4, 5, 6. Calculate the average (mean) rainfall.

4. The set of numbers $\{8, 5, 4, 7, 6, x\}$ has mean 7. Find the value of x, and the median of the set of numbers.

5. The heights, in centimetres, of seven pupils are 126, 124, 117, 129, 132, 120, 127. Find their mean height.

6. Write down three different numbers whose mean is 6 and median 5.

7. On a box of 'Signet' matches is printed "Average contents 50 matches". The contents of 100 boxes were counted with the results:

Number of matches in a box	47	48	49	50	51	52
Number of boxes	2	6	28	45	15	4

(a) Calculate (i) the mean and (ii) the mode number of matches per box.
(b) A box is taken at random from the sample. Find the probability that it will contain at least 50 matches.

STATISTICS
7.2 Charts

7

Piechart and barchart

A **piechart** represents data by the sectors of a circle, or "pieces of pie". The **angle** of a sector is proportional to the **frequency** of the corresponding class of data. A **barchart** represents data by a series of parallel bars. The **height** of a bar is proportional to the **frequency** of the corresponding class of data.

Example Julie's take home pay is £90 per week. Out of this, she plans to spend £20 on rent, £25 on food, £15 on running her car and £10 on entertainment. Of the rest she will spend half on clothes and save the other half. Express this information in a piechart.

The table shows how the £90 is shared out. After rent, food, car and entertainment the remainder is £90 − £20 − £25 − £15 − £10 = £20, so £10 is spent on clothes and she saves £10. The total angle in a piechart is 360°, so £90 : 360° = £1 : 4° and the angle column is found by multiplying the costs column by 4.

Total	£90	360°
rent	£20	80°
food	£25	100°
car	£15	60°
ent'nt	£10	40°
clothes	£10	40°
save	£10	40°

The piechart is drawn in Figure 1, below.

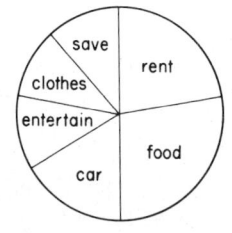

Fig. 1 Piechart showing Julie's expenditure.

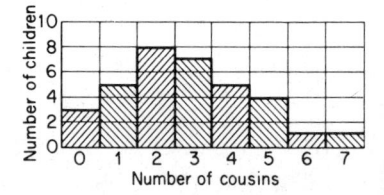

Fig. 2 Barchart showing number of cousins of children in class 2T.

Example The barchart in Figure 2 shows the number of cousins of the children of class 2T. (a) How many children have more than 4 cousins? (b) Write down the mode. (c) Calculate the mean number of cousins per child for the children in class 2T.

(a) The number of children with more than 4 cousins is $4 + 1 + 1 = 6$.
(b) The mode, the tallest bar, is **2 cousins**.
(c) The total number of cousins is $(0 \times 3) + (1 \times 5) + (2 \times 8) + (3 \times 7) + (4 \times 5) + (5 \times 4) + (6 \times 1) + (7 \times 1) = 95$. The total number of children is $3 + 5 + 8 + 7 + 5 + 4 + 1 + 1 = 34$, so the mean is $95/34 = $ **2.8 (2SF)** cousins per child.

Charts 136

SAMPLE QUESTIONS

1 The pie chart in Figure 3 shows how a Sports Centre is being used. There are 8 people weight lifting. How many people are using the centre? How many are playing either squash or table tennis?

2 The numbers of children in each of 18 families is as follows:

Number of children in family	0	1	2	3	4
Number of families	4	3	6	4	1

(*a*) Draw a piechart to show this information. (*b*) Draw a barchart to show this information. (*c*) Write down the modal number of children per family. (*d*) Calculate the mean number of children per family.

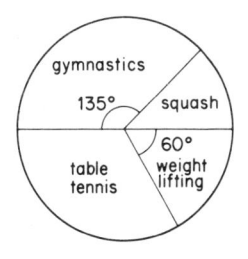

Fig. 3 Use of Sports Centre.

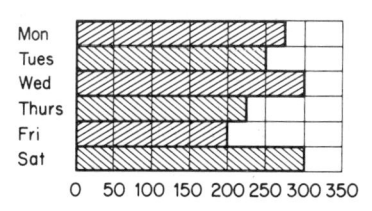

Fig. 4 Number of stamps sold each day.

3 Figure 4 shows the number of first class stamps sold each day in a post office during one week in December. (*a*) How many stamps were sold on the Thursday? (*b*) How many stamps were sold during the week? (*c*) What is the daily average (mean) number of first class stamps sold during that week? (*d*) Use this information to estimate the value of the total number of first class stamps sold in one year (52 weeks) if a first class stamp costs 18p. (*e*) State one reason why your estimate may be inaccurate.

4 In a piechart showing Grant and Ellie's monthly household spending, the angle of the sector representing food is 48°. Their total monthly household spending is £240, find how much they spend on food.

5 Draw an accurate piechart representing the following information about the pocket money received by the children in a class.

Pocket money received	50p	60p	70p	80p	90p	£1
Number of children	4	10	7	2	5	8

How many children received at least 70p per week pocket money? Calculate the average (mean) pocket money received.

STATISTICS
7.3 Pictorial Information   7

Pictograms

A **pictogram**, often used in the media, is a pictorial way of representing data. Each picture stands for a given number, the **scale number**, and parts of a picture may be used.

Example Figure 1 shows the number of TV sets in four towns. Each symbol represents 100 sets. How many TV sets are there in: (a) Brinksley, (b) Sadborough? How many more TV sets are there in Upton than Grimesport?

Brinksley	♇ ♇ ♇ ♇ ♇ ♇ ♇ ♇
Grimesport	♇ ♇ ♇ ♇ ♇ ╷
Upton	♇ ♇ ♇ ♇ ♇ ♇ ♇ ♇ ♇ ♇ ♇ ♇ ♇
Sadborough	♇ ♇ ♇ ♇ ╷

Fig. 1 Pictogram: TV sets in Rorkshire.

(a) There are 8 symbols opposite Brinksley, representing $8 \times 100 = $ **800** sets.
(b) There are $4\frac{1}{2}$ symbols opposite Sadborough, $4\frac{1}{2} \times 100 = $ **450** sets. There are $13 \times 100 - 5\frac{1}{2} \times 100 = 1300 - 550 = $ **750** more sets.

Misleading scale pictures

The size of a picture is often used to indicate the frequencies of the data. In a 2-dimensional picture the frequency should be represented by the **area** of the figure and not by its **linear** dimensions.

*****Example** State why the advertising in Figure 2 is misleading.

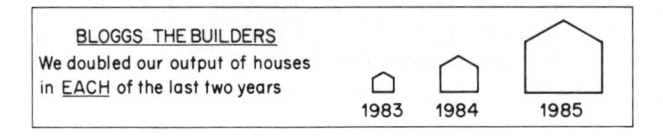

Fig. 2 Misleading statistics.

The pictures are drawn with double the linear dimensions each year. This means that the area is multiplied by 4 each year, which gives a misleading view of the growth in the building of houses.

Pictorial Information 138

*Scatter diagrams

When one quantity y, varies with the value of a second quantity x, a **scatter diagram** is used to draw the graph of the function $y = f(x)$. The values, results of measurements, may not lie exactly on a smooth curve, so a **curve of best fit** is drawn by eye to fit the points as closely as possible. If the function is linear, the **line of best fit** may be drawn using a transparent ruler and can then be used to predict a value of y for a given value of x.

***Example** The temperature of a beaker of liquid is measured over a period of time with the following results.

Time (hours)	1100	1102	1104	1106	1108	1110	1114	1116
Temperature (°C)	95	89	80	76	67	60	49	39

Plot these points on a scatter diagram and draw a line of best fit. Use this to estimate the temperature of the liquid at (*a*) 1109, (*b*) 1120.

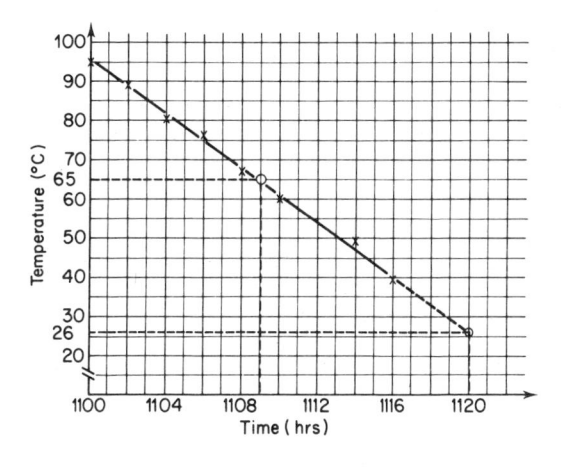

Fig. 3 The cooling of a beaker of liquid.

Figure 3 shows the points plotted and a line of best fit drawn by eye.
From this line the estimated temperatures may be read.
(*a*) At 1109, **65°C**, (*b*) at 1120, **26°C**, BUT this is a very doubtful estimate because this is outside the range of the data. In reality the temperature may well not continue to fall linearly as it nears room temperature.

Pictorial Information

SAMPLE QUESTIONS

1 Figure 4 shows the number of parcel transported by a carrier firm over the last four years. Each small parcel represents 1000 parcels.

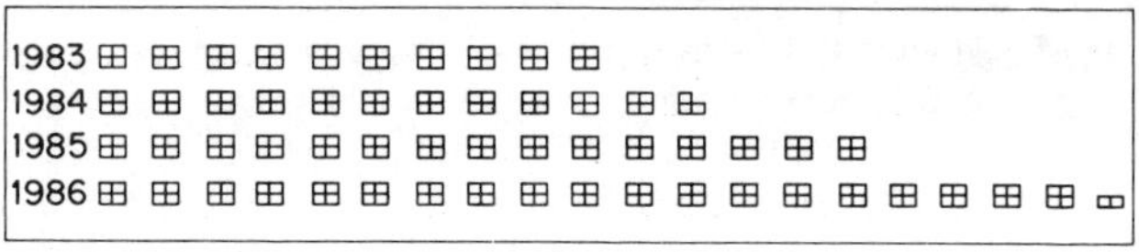

```
1983 ⊞ ⊞ ⊞ ⊞ ⊞ ⊞ ⊞ ⊞ ⊞ ⊞
1984 ⊞ ⊞ ⊞ ⊞ ⊞ ⊞ ⊞ ⊞ ⊞ ⊞ ⊞ ⊞ ⊟
1985 ⊞ ⊞ ⊞ ⊞ ⊞ ⊞ ⊞ ⊞ ⊞ ⊞ ⊞ ⊞ ⊞ ⊞ ⊞ ⊞ ⊞
1986 ⊞ ⊞ ⊞ ⊞ ⊞ ⊞ ⊞ ⊞ ⊞ ⊞ ⊞ ⊞ ⊞ ⊞ ⊞ ⊞ ⊞ ⊞ ⊞ ⊞ ⊞ ⊟
```

Fig. 4 Parcels carried during the last four years.

(a) State the number of parcels carried in 1984. (b) Calculate the percentage increase in the number of parcels carried between 1983 and 1986. (c) How many small parcels will need to be drawn for 1987 if the number of parcels carried falls 4750 below the 1986 figure.

2 State one way in which you think the information given in Figure 5 may be misleading:

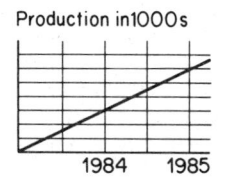

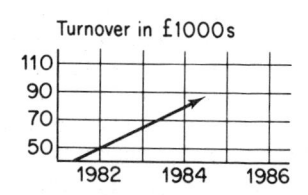

(a) Production has doubled during 1984.　　(b) In 1986 the turnover will be £110 000.

Fig. 5

*3 A manufacturer represents the increased production of refrigerators as shown in Figure 6. If production in 1983 is 1600 refrigerators per year, what production does the diagram represent in 1984 and 1985?

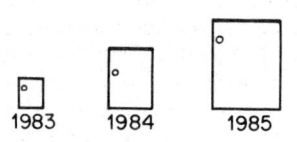

Fig. 6 Production of refrigerators. 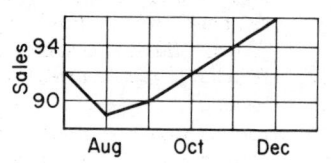　　**Fig. 7** Sales of home computers

4 The statement 'After a poor performance during August, sales have increased dramatically in the months before Christmas' was printed beside the chart drawn in Figure 7. Explain how the chart has been drawn to make this statement appear to be true.

*5 The table show the heights and weights of children being examined at a clinic.

Height (cm)	75	61	86	75	68	91	71
Weight (10 kg)	10.2	5.1	12.4	9.0	6.8	14.0	7.5

Plot these values on a scatter diagram and draw the line of best fit. From this line, find: (a) the expected weight of a child of height 80 cm, (b) the expected height of a child who weighs 11.3 kg.

*6 Weights are hung on a piece of wire and the length measured giving the following table of results.

Weight (kg)	1	2	4	5	8	10	12
Length (cm)	32	32.4	33	33.4	34.6	35.1	35.7

Plot these results on a scatter diagram and draw the line of best fit. From your line estimate: (a) the unstretched length of the wire, (b) the length of the wire when a weight of 6 kg is used, (c) the weight used to stretch the wire to a length of 35 cm. Explain why it is inappropriate to use your line to estimate the length of the wire for a weight of 15 kg.

*7 Plot the data given in the table below on a scatter diagram.

Measurements taken from 12 boys aged 15.

Waist (cm)	71	68	74	79	66	69	76	67	72	76	75	65
Inside leg (in)	30	29	31	33	28	29	32	28	31	33	32	27

Draw the line of best fit through the points by eye. Use your line to estimate: (a) the inside leg measurement for a boy with a waist of 73 cm, (b) the waist measurement for a boy with an inside leg of 34 in. A trouser manufacturer uses the formula $L = W/2 - 4$ to calculate the inside leg length L for trousers of waist W. Draw the line given by this formula on your scatter diagram. State with reasons whether you expect trousers made to this formula to be long or short in the leg for these 12 boys?

STATISTICS
7.4 Grouped Data

7

When there are many different values in the data, it is preferable to **group** values together to form between 5 and 10 groups of data. The **value for the group** is taken as the **mid-point** of the range of values in each group, as demonstrated in the next example.

*Example The following marks are obtained by 100 candidates:

```
21 65 76 88 49 60 38 79 47 83 64 54 67 53 79 52 24 19
54 23 87 40 69 96  7 36 59 76 33 52 68 16 59 61 38 59
13 80 78 85 35 72 46 59 91 37 54 24 87 45 29 78 73 57
49 43 69 55 72 60 78 38 52 61 60 68 37 77 48 60 80 38
73 79  5 77 49 70 51 98 47 65 89 38 47 54 56 77 39 66
43 69 72 40 79 69 74 80 48 44
```

(a) Tally these marks in groups 0–9, 10–19, . . . , 90–99, and produce a **grouped frequency distribution**.
For this distribution find: (b) the mode, (c) the median, (d) the mean.

Group range	Tally table	Frequency	Group value	Freq. × group value
0–9	11	2	4.5	9
10–19	111	3	14.5	43.5
20–29	⊦⊦⊦⊦	5	24.5	122.5
30–39	⊦⊦⊦⊦ ⊦⊦⊦⊦ 1	11	34.5	379.5
40–49	⊦⊦⊦⊦ ⊦⊦⊦⊦ ⊦⊦⊦⊦	15	44.5	667.5
50–59	⊦⊦⊦⊦ ⊦⊦⊦⊦ ⊦⊦⊦⊦ 1	16	54.5	872
60–69	⊦⊦⊦⊦ ⊦⊦⊦⊦ ⊦⊦⊦⊦ 11	17	64.5	1096.5
70–79	⊦⊦⊦⊦ ⊦⊦⊦⊦ ⊦⊦⊦⊦ 1111	19	74.5	1415.5
80–89	⊦⊦⊦⊦ 1111	9	84.5	760.5
90–99	111	3	94.5	283.5
	Total	100	Total	5650

(a) Tally the group ranges by going through the marks in turn and add them up to complete the frequency column. Total this column to check that all 100 marks have been counted.
(b) The mode is in the 70–79 group, so the modal value is **74.5**.
(c) With 100 marks the middle mark is between the 50th and 51st, which both occur in the 50–59 group, counting from either end, so the median mark is **54.5**.
(d) The mean mark is the sum of the products (group value × frequency), calculated in the last column, divided by 100, which is 5650/100 = **56.5**.

*Histograms and continuous data

Grouped data may be shown in a **histogram**, which is like a barchart except that the frequency is indicated by the **area** of the bar instead of the height.

The vertical scale is therefore frequency per unit horizontal scale.

In the examples used so far the data has been **discrete**, the values often being integers. Other data may be **continuous** where the values are any real numbers within some range. When such data is grouped the values in a group must include one end of the group range but not the other. For instance, for lengths x cm, the groups might be

$$0 < x \leqslant 10, \quad 10 < x \leqslant 20, \quad 20 < x \leqslant 30, \ldots$$

***Example** The distances, in kilometres, travelled to school by the 36 pupils in one class are (rounded to the nearest 0.1 km):

4.0 2.2 8.8 3.3 4.1 3.9 0.3 4.4 1.5 3.2 3.8 9.5
2.9 7.5 3.0 3.0 0.8 4.7 4.7 5.6 2.2 3.8 4.4 3.9
2.0 5.0 4.6 3.3 2.5 6.5 4.9 2.5 6.0 2.4 3.2 1.9

Represent this data in a histogram using the group ranges: less than 2 km, 2 km and less than 3 km, 3 km and less than 4 km, 4 km and less than 5 km, 5 km and less than 10 km.

The distances are tallied for the given ranges and the frequencies totalled, in Figure 1(a). The last column, frequency per km, is then used to draw the histogram (Figure 1(b)).

Group range	Tally	Freq.	Freq. per km
$x < 2$	1111	4	$4/2 = 2$
$2 \leqslant x < 3$	⊞ 11	7	$7/1 = 7$
$3 \leqslant x < 4$	⊞ ⊞	10	$10/1 = 10$
$4 \leqslant x < 5$	⊞ 111	8	$8/1 = 8$
$5 \leqslant x < 10$	⊞ 11	7	$7/5 = 1.4$
	Total	36	

Fig. 1 (a) Continuous data frequency table

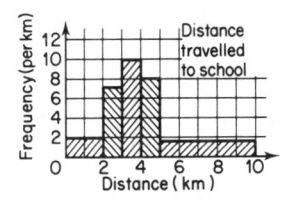

(b) Histogram

*Measures of spread

The difference between the largest and smallest values is called the **range**. For greater sophistication the **upper and lower quartiles** are used. These are $\frac{1}{4}$ and $\frac{3}{4}$ of the way along the distribution when it is ordered (just as the median is $\frac{1}{2}$ way along). The difference between these quartiles is called the **inter-quartile range**.

*Cumulative frequency distribution

For grouped distributions, the calculation of the median and quartiles is eased by producing a **cumulative frequency distribution**. This gives the total number of values less than or equal to the value considered, and is found by adding up all the frequencies obtained so far. A **cumulative frequency diagram** may now be drawn and the median and quartiles marked.

***Example** The number of people staying in a Youth Hostel on each of 60 successive nights is tabulated below. Complete the cumulative frequency column given in Figure 2, and draw a cumulative frequency diagram. Mark the median and quartiles and state the inter-quartile range. The Youth Hostel is considered to be running economically when more than 30 people are in residence. Based on these nights what fraction of the time is the Hostel running uneconomically.

Number of people	Frequency	Cumulative frequency
0–10	1	1
11–15	3	4
16–20	2	6
21–25	4	
26–30	5	
31–35	9	
36–40	10	
41–50	17	
51–60	6	
61–70	3	
Total	60	

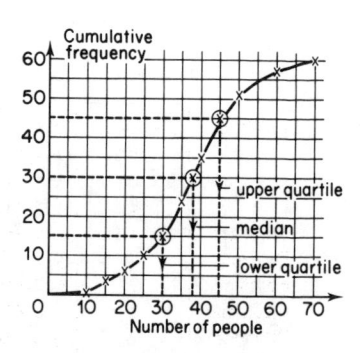

Fig. 2 (a) Cumulative frequency table   (b) Cumulative frequency diagram

The cumulative frequency column is found by adding on each group frequency in turn: **1 4 6 10 15 24 34 51 57 60**.

These points are plotted as shown in Figure 2(b) and joined by a smooth curve. The median and quartiles are marked at the 15th, 30th and 45th nights and give **30, 38** and **45** people, so the inter-quartile range is $45 - 30 = $ **15** people.

For 15 nights out of 60 the Hostel had 30 or fewer people staying, so it is running uneconomically for $\frac{1}{4}$ of the time.

SAMPLE QUESTIONS

***1** The number of words in 60 sentences taken from the *Times* editorials is shown in the table below. Group the data into sentences with 0–5, 6–10, 11–15, 16–20, ..., words, and calculate the mean number of words per sentence.

12	23	14	18	9	21	27	14	19	29	13	16	26	33	14
32	25	12	7	18	22	28	18	26	18	11	24	29	15	9
35	23	27	17	16	21	28	14	26	14	13	8	23	26	13
28	19	20	22	14	17	30	11	10	27	16	14	29	14	17

***2** Last season the school first eleven scored the following number of goals in 48 matches.

2	1	2	0	3	1	0	2	0	4	1	0	3	0	4	1
3	0	1	1	2	2	2	2	3	2	5	3	0	2	5	0
2	0	1	0	1	0	1	0	3	0	0	3	1	2	1	2

Construct a cumulative frequency table of the numbers of goals scored. Find: (a) the range, (b) the mode, (c) the median, (d) the inter-quartile range, (e) the mean, of this distribution.

3 The following advert for staff was placed by the heating firm Cosywarm. The firms annual wages are given in the table for the present staff.

> Cosywarm Central Heating
> More staff needed for
> an expanding business
> AVERAGE WAGES over £10 000
> per year. Apply now!

Position	Number employed	Annual salary
Director	1	£30 000
Manager	1	£24 000
Designer	2	£19 000
Foreman	1	£12 000
Electrician	4	£8 000
Plumber	9	£6 500

Find for the annual wages in the firm: (a) the mode, (b) the median, (c) the mean. Use your answers to state why it might be said that the advertisement is misleading.

***4** Membership of a youth club is confined to young people aged over 12 and under 20. The leader surveys the ages of the current members with the following results:

Age last birthday	12	13	14	15	16 and over
Number of members	6	9	13	11	8

Draw a histogram of these results, using a horizontal scale for ages 12 to 20 and a vertical scale for 'Frequency per one-year interval.'

PROBABILITY
8.1 Taking a Chance

8

Equally likely outcomes
When a normal coin is tossed, it will fall either **heads** (H) or **tails** (T). The two **outcomes** are **equally likely**, so the **probability** of getting H is $\frac{1}{2}$, see Figure 1. When this experiment is repeated a great number of times with an **unbiased** coin, approximately half the outcomes will be H and half will be T.

Outcomes	H	T
Probability	$\frac{1}{2}$	$\frac{1}{2}$

Outcomes	1	2	3	4	5	6
Probability	$\frac{1}{6}$	$\frac{1}{6}$	$\frac{1}{6}$	$\frac{1}{6}$	$\frac{1}{6}$	$\frac{1}{6}$

Fig. 1 Tossing a coin. **Fig. 2** Throwing a die.

When a normal 6-sided die is thrown, there are 6 **equally likely outcomes**. The probability of the **event** 'throwing a 2' is $\frac{1}{6}$, see Figure 2.
The probability of the event '**not** throwing a 2' is $\frac{5}{6}$, since there are 5 outcomes which do **not** give 2.

$$\text{Probability of an event} = \frac{\text{Number of outcomes leading to the event}}{\text{Total number of equally likely outcomes}}$$

A probability is a **fraction** between 0 and 1, 0 if the event **never** occurs, and 1 if the event **always occurs**.

When an object is chosen **at random** from n objects the choice of any particular object is equally likely, and the probability is $\dfrac{1}{n}$.

Example A letter is chosen at random from the word MANCHESTER. Find the probability that it is; (a) the letter E, (b) not a vowel.

(a) There are 10 equally likely outcomes, 2 of which lead to the letter E, so the probability of choosing E is $\frac{2}{10}$, or $\frac{1}{5}$, cancelling down the fraction.
(b) There are 3 vowels, A, E, E, in the word, so the probability of 'choosing a vowel' is $\frac{3}{10}$, or 0.3. The probability of 'not choosing a vowel' is therefore 0.7, or $\frac{7}{10}$, since these two probabilities must add up to 1. There are 7 consonants, 'non-vowels', in the given word.

Example Five cards, labelled P, Q, R, S, T, are shuffled and two cards are selected at random. (a) List all the possible outcomes. (b) Find the probability that the pair labelled Q and S are chosen. (c) What is the probability that the card labelled R is one of the pair chosen?

(a) List the possible outcomes systematically to make sure none are missed. List all those containing P, then all those with Q, but not P and so on.
 PQ PR PS PT QR QS QT RS RT ST — 10 equally likely events.
(b) The probability of choosing the pair QS is $\frac{1}{10}$.
(c) 4 outcomes *PR, QR, RS* and *RT* contain R, so the probability is $\frac{4}{10} = \frac{2}{5}$.

Taking a Chance 146

SAMPLE QUESTIONS

1. The probability of Helen winning a race is $\frac{1}{3}$. State the probability that she does not win the race.
2. The 9 letters LIVERPOOL are put into a box and one is drawn at random. Find the probability of choosing (a) an L, (b) a consonant.
3. The probability that Robin will miss the target when he shoots a crossbow is 0.15. Find the probability that he will hit the target.
4. Find the probability of getting: (a) an odd number, (b) a square number, on one throw of an unbiased die.
5. One pupil is chosen at random from a class containing 17 boys and 11 girls. Find the probability that a girl is chosen.
6. A normal coin is tossed 4 times and the first 3 times it comes down as 'heads'. What is the probability that the last throw will be 'heads'.
7. A pack of 52 playing cards contains 13 cards, A, 2–10, J, Q, K of each of 4 suits; clubs, diamonds, hearts, spades. Find the probability that a card chosen at random is: (a) a heart, (b) the Queen of Hearts, (c) an Ace.
8. Figure 3 shows a game. The pointer is spun and stops, pointing at a sum of money, in pence, which the player wins. Find the probability that: (a) the player wins nothing, (b) the player wins 5p or more.
9. Complete the list of outcomes when a coin is tossed three times: HHH HHT HTH ... ... TTT. Find the probability that the result is two heads and one tail in any order.

Fig. 3 Fairground game.   **Fig. 4** History and Geography.

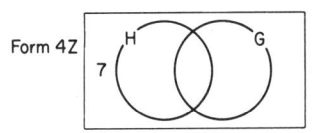

*10. There are 32 pupils in 4Z. Nineteen study History, 15 Geography and 7 neither. Complete the Venn diagram in Figure 4 showing the number of pupils studying History (H) or Geography (G). Find the probability that a pupil chosen at random from 4Z studies both History and Geography.
11. Elliot has a bag of chocolates. There are 14 plain chocolates, 9 with hard centres and 5 soft centred, and 6 milk chocolates, half of which are hard and half soft centred. He offers Letty a chocolate and she picks one at random from the bag. What is the probability that she picks: (a) a plain, (b) a milk with a soft centre, (c) a soft centre.
12. There are 80 red and 20 white counters mixed in bag. One counter is drawn at random. What is the probability that it is red? The first counter is white and is not replaced. What is the probability that a second counter drawn is red?

PROBABILITY
8.2 Combined Events

8

When two events, each with a set of equally likely outcomes, occur together, it is convenient to represent the **sample space** of possible outcomes on a two dimensional coordinate grid as shown in the next example.

Example Two dice are thrown at the same time; a red one with six equal faces numbered 1 to 6 and a blue one with four equal faces labelled 2 4 6 8. Copy Figure 1 which shows the sample space of outcomes. (*a*) Draw a ring round each outcome where the total score equals 8 or 9 and hence state the probability that this event occurs. (*b*) Draw a cross on each outcome where the difference between the score on the two dice is 3 and hence state the probability of this event. (*c*) State the probability that either of these two events occurs.

	red die								red die				
	1	2	3	4	5	6			1	2	3	4	5	6
blue 2	.	.	.	.	.	.	blue 2	.	.	.	.	X	O
die 4	.	.	.	.	.	.	die 4	X	.	.	O	O	.
6	.	.	.	.	.	.	6	.	O	(X)	.	.	.
8	.	.	.	.	.	.	8	O	.	.	.	X	.

Fig. 1 Sample space for two dice. **Fig. 2** O = total 8 or 9, X = differ by 3.

Figure 1 shows the sample space as a set of 24 coordinate pairs, (blue, red), each dot representing one of the possible outcomes.

(*a*) The points (2, 6), (4, 4), (4, 5), (6, 2), (6, 3) and (8, 1), 'ringed' in Figure 2, each represent an outcome with total score 8 or 9. The probability of this event is $\frac{6}{24} = \frac{1}{4}$.

(*b*) The points (2, 5), (4, 1), (6, 3) and (8, 5), 'crossed' in Figure 2 each represent an outcome with scores on the two dice differing by 3. The probability of this event is $\frac{4}{24} = \frac{1}{6}$.

(*c*) The points where either of these two events occur are the set of 'ringed' or 'crossed' points, 9 (= 6 + 4 − 1) in number since (6, 3) occurs in both events. The probability of this combined event is $\frac{9}{24} = \frac{3}{8}$.

Conditional probability

Sometimes two events follow one another and the set of outcomes of the second event depends on the outcome of the first event. In this situation a **tree diagram** can be used to show the possible combined events and their probabilities.

Example A bag contains 3 red counters and 5 green counters. Two are taken out at random. Complete the tree diagram given in Figure 3 showing the 4 possible events and their probabilities. Find the probability of: (*a*) the two counters being the same colour, (*b*) one of each colour, (*c*) at least one of red counter being drawn.

Combined Events 148

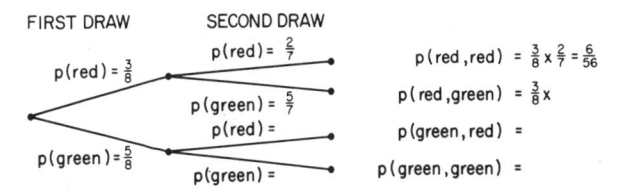

```
FIRST DRAW        SECOND DRAW
                  p(red) = 2/7                p(red, red) = 3/8 × 2/7 = 6/56
p(red) = 3/8
                  p(green) = 5/7              p(red, green) = 3/8 ×
                  p(red) =                    p(green, red) =
p(green) = 5/8
                  p(green) =                  p(green, green) =
```

Fig. 3 Tree diagram, selection of counters.

Following the upper branch means that the first counter drawn is red, with probability $\frac{3}{8}$, and there are then only 7 counters left for the second draw. The probability of this second counter being red is $\frac{2}{7}$, and the probability of a green second counter is $\frac{5}{7}$. Following the lower branch means the first counter drawn is green, probability $\frac{5}{8}$. The probability of the second counter being red is now $\frac{3}{7}$, and being green is $\frac{4}{7}$.

Multiplying along the branches the combined events have probabilities:

$$p(\text{red, red}) = \tfrac{3}{8} \times \tfrac{2}{7} = \tfrac{6}{56}$$
$$p(\text{red, green}) = \tfrac{3}{8} \times \tfrac{5}{7} = \tfrac{15}{56}$$
$$p(\text{green, red}) = \tfrac{5}{8} \times \tfrac{3}{7} = \tfrac{15}{56}$$
$$p(\text{green, green}) = \tfrac{5}{8} \times \tfrac{4}{7} = \tfrac{20}{56}$$

Notice how this final column adds up to $\frac{56}{56} = 1$, since one of these four combined events must always occur.

(a) The probability of the two counters being the same colour is
p(red, red) or p(green, green) = $\frac{6}{56} + \frac{20}{56} = \frac{26}{56} = \frac{13}{28}$.
(b) p(one of each) is: p(red, green) or p(green, red) = $2 \times \frac{15}{56} = \frac{30}{56} = \frac{15}{28}$.
(c) p(at least 1 red counter) is
$1 - p(\text{no red counters}) = 1 - p(\text{green, green})$
so the probability of at least 1 red is $1 - \frac{20}{56} = \frac{36}{56} = \frac{9}{14}$.

***Example** Ahmed and Bella play a game, taking a ball at random from a box which holds 4 white and 2 red balls initially, the winner being the first person to pick a white ball. When Ahmed picks the first ball find the probability that: (*a*) the first ball he picks is white, (*b*) Bella picks a ball, (*c*) Bella wins.

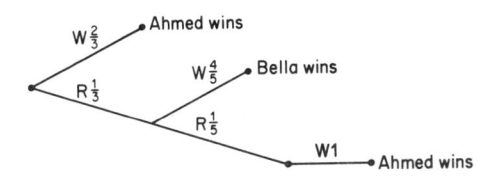

```
         W 2/3  → Ahmed wins
        ╱
       ╱    W 4/5  → Bella wins
R 1/3 ╲    ╱
       ╲  ╱
        R 1/5
              ╲  W1
               ╲    → Ahmed wins
```

Fig. 4 Ahmed and Bella's game.

Combined Events

Figure 4 shows the tree diagram for Ahmed starting.
(a) The probability Ahmed picks a white is $\frac{4}{6} = \frac{2}{3}$.
(b) Bella only picks a ball if Ahmed takes a red on his first pick, so the probability is $\frac{2}{6} = \frac{1}{3}$.
(c) Bella only wins if she takes a white on her first go, since if she takes the other red Ahmed is bound to take a white on his second go. Therefore the probability of Bella winning is $\frac{2}{6} \times \frac{4}{5} = \frac{4}{15}$.

The examples above contain instances where probabilities are added and others where probabilities are multiplied together.

*Consider two events called A and B, then:

$$\boxed{p(A \text{ OR } B) = p(A) + p(B) - p(A \text{ AND } B)}.$$

$$\boxed{p(A \text{ AND } B) = p(A) \times p(B \text{ given } A)},$$

where p(B given A) means the probability of 'B given that A has happened already'.

In some cases p(B given A) is the same as p(B) in which case events A and B are said to be **independent**, and then

$$\boxed{p(A \text{ AND } B) = p(A) \times p(B)}.$$

In some cases A and B can't occur together, so p(A AND B) = 0, in which case events A and B are said to be **mutually exclusive**, and then

$$\boxed{p(A \text{ OR } B) = p(A) + p(B)}.$$

***Example** Calculate the probability of obtaining: (a) a six when a fair die is thrown, (b) a head when a fair coin is tossed, (c) a six and a head when the die and the coin are thrown together, (d) a six or a head (or both) when they are again thrown together.

Let S be the event 'a 6 on the die' and H be 'a head on the coin'.
(a) $p(S) = \frac{1}{6}$,
(b) $p(H) = \frac{1}{2}$,
(c) $p(S \text{ AND } H) = p(S) \times p(H) = \frac{1}{6} \times \frac{1}{2} = \frac{1}{12}$, since tossing the coin and throwing the die are independent.
(d) $p(S \text{ OR } H) = p(S) + p(H) - p(S \text{ AND } H) = \frac{1}{6} + \frac{1}{2} - \frac{1}{12} = \frac{7}{12}$.

Combined Events 150

SAMPLE QUESTIONS
1. Two dice are thrown together. Find the probability that the scores add up to 10. Find the probability that the scores differ by at least 3.
2. In a box of pencils, 27 are sharp and 9 are blunt. Drawn a tree diagram to find the probability that when two pencils are taken from the box without replacing: (*a*) both are blunt, (*b*) one is sharp and one blunt.
3. In a board game a die is thrown and a counter then moved along a path. If the score on the die is odd the counter is moved forward by that score, whilst if the score is even the counter is moved backwards by half the score. Find the probability that after two moves: (*a*) the counter is in its original position, (*b*) the counter has moved forwards.
4. A bag contains 6 red and 8 blue beads. One bead is taken from the bag. (*a*) Find the probability that it is red. The bead is replaced and 2 beads are then taken from the bag. Find the probability that they are: (*b*) both red, (*c*) both blue.
5. Each day the probability that Bill is early for school is 1/12, the probability that he is late is 1/4, and otherwise he is on time. Find (*a*) how many times he is likely to arrive on time out of a term of twelve 5-day weeks, (*b*) the probability that he is late next Monday and Tuesday.
*6. Judith's driving instructor tells her that she has a probability of 4/5 of passing the test each time she takes it. Assuming he is correct, state as a fraction, the probability that she will: (*a*) fail at her first attempt, (*b*) pass at the third attempt having failed the first two.
*7. I throw an unbiased die. State the probability of getting a five. I now throw the die again and then toss a coin. Find the probability of getting either a five or a head or both.
*8. Two piles of exercise books lie on a table. Pile A contains 4 maths books and 2 physics books; Pile B contains 3 maths and 3 physics books. Two changes are made at random in the following order:
Change 1. A book is taken from Pile A and put in Pile B.
Change 2. A book is removed from Pile B.
Draw a tree diagram showing the probability of each possible outcome. Find the probability that a maths book is removed from Pile B.
*9. The names of pupils are drawn from a hat in order to decide which of the 10 girls and 6 boys should read the next extract from a play. (*a*) Find the probability that the first name drawn is a boy. (*b*) Find the probability that the first two names drawn are both girls.
*10. Kieran plays draughts with Maureen. If he starts, the probability that he wins is 1/3 and they draw is 1/6. If she starts, the probability of her winning is 1/6 and losing is 1/2. (*a*) If Kieran starts what is the probability that he loses. (*b*) If they play two games, taking turns to start, find the probability that Kieran, (i) wins, (ii) loses, both games.

TRANSFORMATIONS
9.1 Enlarge it   9

Similarity and enlargement

Figure 1 shows a photograph, a double size **enlargement**, and a half size reduction. The three rectangular shapes are mathematically **similar**, meaning they have the **same shape** but a **different size**. The **scale factor** of the enlargement is **2**, so any length in the photograph will have a similar length in the enlargement which is twice as long. The reduction is an example of a scale factor $\frac{1}{2}$ enlargement.

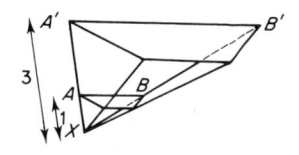

Fig. 1   Photo, $\times 2$ $\times \frac{1}{2}$ enlargements.   **Fig. 2**   Scale factor 3 enlargement.

Figure 2 shows a shape which has been enlarged **scale factor 3**, using the point X as the **centre of enlargement**.

Corresponding lengths AB and $A'B'$ are in the ratio $1:3$, so $3 \times AB = A'B'$, but the shape was produced by extending XA to XA', since $3 \times XA = XA'$.

Example   The triangle ABC with vertices $A(1, 2)$, $B(-1, 0)$, $C(1, -1)$ is transformed by an enlargement **E**, scale factor 2, centre $(0, 0)$, to the triangle $A'B'C'$.
(a)   Draw the triangles and state the coordinates of A', B' and C'.

The translation **T** given by the vector $\binom{1}{2}$ maps $A'B'C'$ to $A''B''C''$.
(b)   Draw triangle $A''B''C''$ on your diagram.
(c)   Describe fully a transformation mapping ABC onto $A''B''C''$.

(a)   Since **E** has centre $(0, 0)$ and scale factor 2 the coordinates of each point are doubled. So $A' = (2, 4)$, $B' = (-2, 0)$, $C' = (2, -2)$, see Figure 3.
(b)   **Translating** $A'B'C'$ 1 across and 2 up gives $A''B''C''$, see Figure 3.
(c)   Clearly ABC to $A''B''C''$ is an **enlargement scale factor 2**, with centre of enlargement at the point where $A''A$, $B''B$, $C''C$, shown dotted, meet. So $(-1, -2)$ is the **centre** of enlargement.

Enlarge it 152

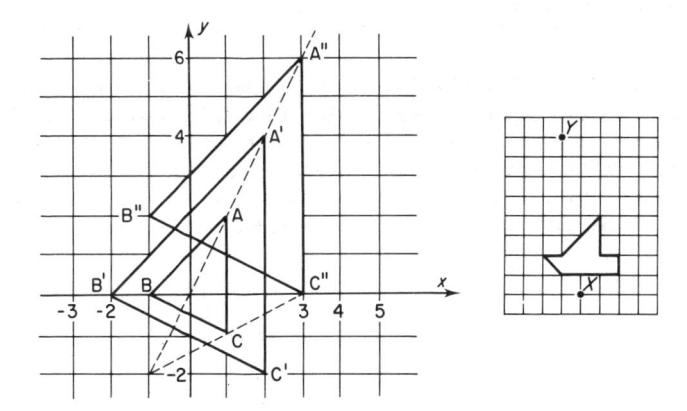

Fig. 3 Enlargement of a triangle.

Fig. 4 Boat shape.

SAMPLE QUESTIONS

1. In Figure 4, enlarge the boat shape: (*a*) scale factor 2 with *X* as centre of enlargement, (*b*) scale factor 0.5 with *Y* as centre of enlargement.
2. (*a*) In Figure 5 draw *F'*, the result of transforming the letter *F* by an enlargement, centre (0, 1), scale factor 2.
   (*b*) *F'* is now transformed by an enlargement, centre (6, 1), scale factor $\frac{1}{2}$, to *F"*.
   (*c*) Describe a single transformation which maps *F* to *F"*.

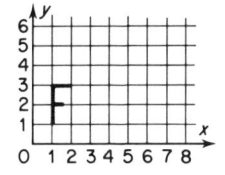

Fig. 5 Transformations of F.

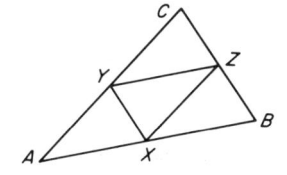

Fig. 6 Triangle transformations.

3. In Figure 6, *X*, *Y*, *Z* are the midpoints of the sides *AB*, *AC*, *BC* of the triangle *ABC*. Describe fully the transformations which map: (*a*) triangle *AXY* onto *YZC*, (*b*) triangle *AXY* onto *ZYX*, (*c*) triangle *AXY* onto *ABC*.
4. The triangle with vertices *P*(2, 6), *Q*(3, 8), *R*(4, 6) is enlarged to *P'*(0, 8), *Q'*(3, 14), *R'*(6, 8). Find the scale factor and centre of enlargement.
5. Draw a regular hexagon *RSTUVW* with sides 3 cm in length. Draw the following enlargements of the hexagon: (*a*) centre *R*, scale factor 3, (*b*) centre *T*, scale factor $\frac{1}{3}$, (*c*) centre *V*, scale factor $1\frac{1}{2}$.

TRANSFORMATIONS
9.2 Translation, reflection, rotation

There are three different types of transformation in which the **distance between points is unaltered**, **translation**, **rotation** and **reflection**. This means that angles between corresponding lines are also unaltered and figures are transformed into **congruent** figures. The properties of these three **isometries** are listed below:

A **translation** shifts every point by the same **vector displacement**.
A **reflection** reflects every point in the given **mirror line**.
A **rotation** rotates every point through a **given angle** about a fixed point, the **centre of rotation**. Rotations through a **positive** angle are taken **anticlockwise**, rotations through a **negative** angle are **clockwise**.

Example Draw the effect of the transformation on the triangle ABC, $A(-1, 1)$, $B(1, 1)$, $C(1, 4)$: (a) translation by the vector $\begin{pmatrix} 3 \\ -2 \end{pmatrix}$, (b) reflection in the mirror line $y = 0$, (c) reflection in mirror line $x + y = 0$, (d) clockwise rotation of $90°$ about the origin $(0, 0)$.

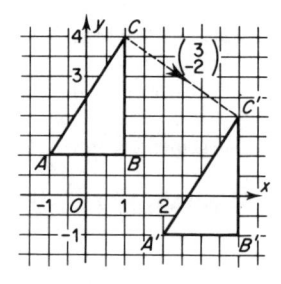

Fig. 1 (a) Translation.

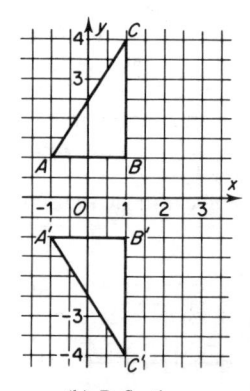

(b) Reflection.

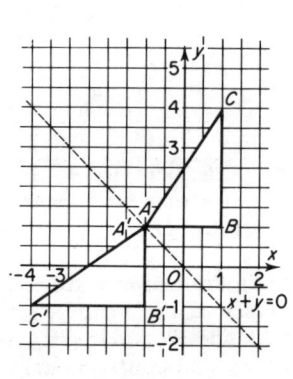

(c) Reflection.

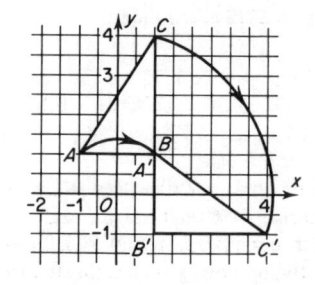

(d) Rotation.

Translation, Reflection, Rotation

The effect of these four transformations is drawn in Figure 1.
(a) The displacements $\overrightarrow{AA'}$ $\overrightarrow{BB'}$, $\overrightarrow{CC'}$ are all 3 across and 2 down.
(b) $y = 0$ is the x-axis, so $A(-1, 1)$ goes to $A'(-1, -1)$.
(c) $y + x = 0$ passes through $(0, 0)$ with gradient -1, so A stays fixed, $B(1, 1)$ goes to $B'(-1, -1)$ and $C(1, 4)$ goes to $C'(-4, -1)$.
(d) Rotate OA 90° clockwise about O to OA' where A' has coordinates $(1, 1)$.

SAMPLE QUESTIONS

1. On Figure 2(a) and (b) draw the reflections of the shape in the mirror line.

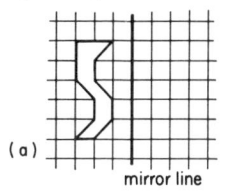

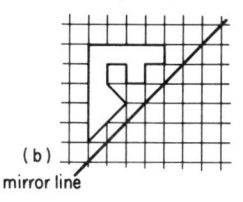

Fig. 2 Reflections.

2. On Figure 3, (a) draw the reflection $A'B'C'$ of the triangle ABC in the line LM, (b) draw the reflection $P'Q'R'$ of the triangle PQR in ST.

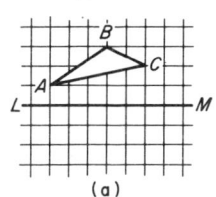

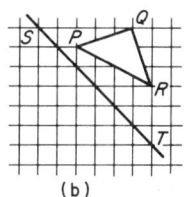

Fig. 3 Reflections.

3. On the grid of Figure 4 draw the new positions of the **Z** shape after: (a) a translation with vector $\binom{7}{2}$, (b) a clockwise rotation through half a turn about **X**, (c) a reflection in the mirror line.

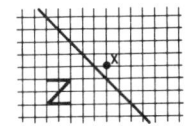

Fig. 4 Z shape. 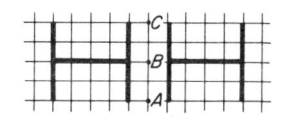   **Fig. 5** H shapes.

4. Describe completely three different transformations which map the lefthand **H** on to the righthand **H** in Figure 5.
5. The triangle ABC, $A(1, 2)$, $B(3, 4)$, $C(4, 3)$ is transformed to $A'B'C'$ under, (a) a translation, (b) a rotation of 180°, where $A'(5, 0)$ is the image of A. Give the coordinates of A', B', and C', in each case.

TRANSFORMATIONS
9.3 *Combined Transformations   9

A transformation **T** can be thought of as a **function** on the points of the plane.
If **T** maps P to P' then we may write $\mathbf{T}(P) = P'$.
Suppose another transformation **S** then maps P' to P'',

$\mathbf{S}(P') = P''$, so $\mathbf{S}(\mathbf{T}(P)) = P''$, written $\mathbf{ST}(P) = P''$,

where **ST** is called the **composition** of the transformations, **T** followed by **S**.
Note the **reversed order** as in composition of functions.

*Example Transformations **S** and **T** are defined as follows:
S is the rotation through half a turn about the origin O, $(0, 0)$,
T is the rotation through half a turn about the point P, $(2, 0)$.
(a) Copy Figure 1 showing the position of triangle A.
(b) Draw and label triangles B and C, where $\mathbf{S}(A) = B$, $\mathbf{T}(B) = C$.
(c) Describe the transformation **TS** which transforms A into C.
(d) Draw and label triangles D and E where $\mathbf{T}(A) = D$, $\mathbf{S}(D) = E$.
(e) Describe the transformation **ST** which transforms A into E.
(f) Describe the relation between the transformations **TS** and **ST**.

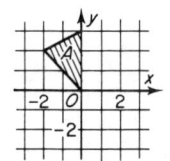

Fig. 1 Triangle A.

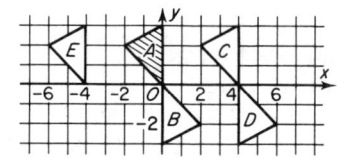

Fig. 2 Triangles B, C, D and E.

(a) and (b) Figure 2 shows triangles A, B and C.
  Triangle B is found by rotating A $180°$ about O.
  Triangle C is found by rotating B $180°$ about P.
(c) The transformation which moves A to C is a translation through 4 units parallel to the x-axis, so **TS** is the translation $\begin{pmatrix} 4 \\ 0 \end{pmatrix}$.
(d) Figure 2 also shows triangles D and E, found by rotating A about $(2, 0)$ and then about $(0, 0)$.
(e) Moving A to E is another translation, -4 units parallel to the x-axis, so the transformation **ST** is the translation $\begin{pmatrix} -4 \\ 0 \end{pmatrix}$.
(f) **ST** is the opposite translation to **TS**, called the **inverse transformation**.

Combined Transformations 156

Inverse transformations

The **identity transformation, I,** leaves every point fixed, so $I(P) = P$, for every point P in the plane.

The **inverse transformation, T^{-1}** of a given transformation **T** is the opposite of **T**. T^{-1} returns every point back to where it was before it was moved by **T**, so:

$T^{-1}(T(P)) = P$, and $T^{-1}T = I = TT^{-1}$.

As seen in the example above, the inverse of a translation is the **negative translation**. The inverse of a rotation is a rotation about the same centre, but through the **negative angle**. However the inverse of a reflection is the **same** reflection, since two reflections in a mirror line return a figure to its starting point.

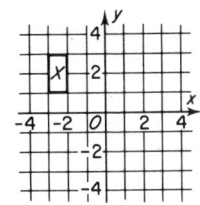

Fig. 3 Rectangle X.

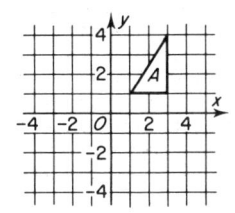

Fig. 4 Triangle A.

SAMPLE QUESTIONS

1. Copy Figure 3, draw and label rectangles Y and Z where:
   (a) Y is the image of rectangle X after a reflection in $y = x$,
   (b) Z is the image of Y after a rotation through $90°$ about $(0, 0)$.
   Describe the single transformation which will transform X into Z.

2. On a copy of Figure 4, draw triangles B, C and D where triangle A is transformed into: (i) B by a half turn about the origin O, (ii) C by a half turn about the point $(3, 1)$, (iii) D by a reflection in the y-axis. Describe a single transformation which maps: (i) B onto C, (ii) B onto D.

*3. **R** is a reflection in the x-axis, **S** is a reflection in the y-axis and **T** is a half turn rotation about the origin. X is a triangle with vertices at points $(4, 2)$, $(3, 5)$ and $(1, 3)$. Draw (a) X, (b) $R(X)$, (c) $SR(X)$, (d) $TR(X)$.
   Give single transformations equivalent to: (e) **SR**, (f) **TR**, (g) **TSR**, (h) **RST**.

*4. Line L has equation $x = 2$, **T** is a rotation of $90°$ about the point $(1, 0)$, and **M** is a reflection in the line $x = y$. Give the equation of the following lines: (a) $T(L)$, (b) $TT(L)$, (c) $T^{-1}(L)$, (d) $M(L)$, (e) $MT(L)$, (f) $TM(L)$.

VECTORS AND MATRICES
10.1 Vectors

10

Translations
Vectors may be regarded as **displacements**. If a vector **a** corresponds to the displacement from A to B then $\mathbf{a} = \overrightarrow{AB}$. The point A is the **start** and B is the **end** of the vector $\overrightarrow{AB}$. The displacement $\overrightarrow{CD}$ is the same as $\overrightarrow{AB}$ if it is in the **same direction** and of the **same magnitude**, length. In this case the vectors are equal, $\overrightarrow{AB} = \overrightarrow{CD}$, and this means $ABDC$ is a parallelogram, see Figure 1(a).

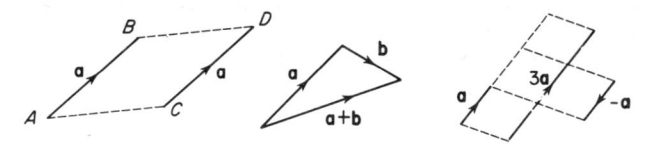

Fig. 1 (a) Equal vectors, (b) Adding vectors, (c) Multiply by a scalar.

Two displacements are added by performing the second displacement after the first and vectors are added in the same way. To add vectors **a** and **b** place the start of **b** at the end of **a** then $\mathbf{a} + \mathbf{b}$ is the vector from the start of **a** to the end of **b**, see Figure 1(b).

A vector **a** multiplied by a scalar (number), k, gives a vector $k\mathbf{a}$ parallel to **a** but k times as long.

When k is positive then the direction of $k\mathbf{a}$ is the same as the direction of **a**, see Figure 1(c) which shows **a** and $3\mathbf{a}$.

When k is negative then the direction of $k\mathbf{a}$ is parallel to **a** but reversed.

When $k = -1$ then $k\mathbf{a} = -\mathbf{a}$, giving the **negative**, or opposite vector, see Figure 1(c). If $\mathbf{a} = \overrightarrow{AB}$ then $-\mathbf{a} = \overrightarrow{BA}$, and **subtraction** of vectors is performed by **adding the negative**.

Example On the diagram, Figure 2, construct vector displacements: (*a*) equal to $\overrightarrow{AB} + \overrightarrow{CD}$, (*b*) equal to $\overrightarrow{AB} - \overrightarrow{CD}$.

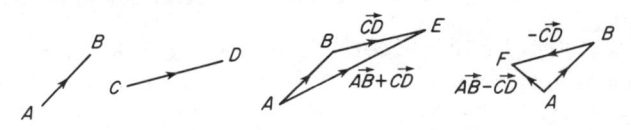

Fig. 2 Vectors $\overrightarrow{AB}$ & $\overrightarrow{CD}$.   **Fig. 3** (a) $\overrightarrow{AB} + \overrightarrow{CD}$, (b) $\overrightarrow{AB} - \overrightarrow{CD}$.

(a) In Figure 3(a) construct the displacement $\overrightarrow{BE}$ equal to $\overrightarrow{CD}$, then $\overrightarrow{BE} = \overrightarrow{CD}$ and $\overrightarrow{AB} + \overrightarrow{CD} = \overrightarrow{AB} + \overrightarrow{BE} = \overrightarrow{AE}$.

(b) In Figure 3(b) let $\overrightarrow{BF} = \overrightarrow{DC} = -\overrightarrow{CD}$, then
$\overrightarrow{AB} - \overrightarrow{CD} = \overrightarrow{AB} + \overrightarrow{BF} = \overrightarrow{AF}$.

Vectors 158

*Example In Figure 4, *ABCDEF* is a regular hexagon with centre *O*. $\vec{AB} = \mathbf{a}$, $\vec{BC} = \mathbf{b}$, $\vec{CD} = \mathbf{c}$ and $\vec{DE} = \mathbf{d}$. (*a*) Write down in terms of **a** and **b**, (i) $\vec{AC}$, (ii) $\vec{AO}$, (iii) $\vec{AD}$. (*b*) Show, (i) $\mathbf{a} + \mathbf{d} = \mathbf{0}$, (ii) $\vec{AB} + \vec{AF} = \mathbf{b}$, (iii) $\mathbf{a} + \mathbf{b} + \mathbf{c} = 2\mathbf{b}$.

(*a*) (i) $\vec{AC} = \vec{AB} + \vec{BC} = \mathbf{a} + \mathbf{b}$. (ii) *AOCB* is a parallelogram so $\vec{AO} = \vec{BC}$ and $\vec{AO} = \mathbf{b}$. (iii) $\vec{AD} = 2\vec{AO}$ so $\vec{AD} = 2\mathbf{b}$.
(*b*) *ABDE* is a rectangle so $\vec{DE} = \vec{BA} = -\vec{AB}$. Therefore $\mathbf{a} + \mathbf{d} = \vec{AB} + \vec{DE} = \vec{AB} - \vec{AB} = \mathbf{0}$. (ii) From two parallelograms $\vec{AO} = \mathbf{b}$ and $\vec{AF} = \vec{BO}$, so $\vec{AB} + \vec{AF} = \vec{AB} + \vec{BO} = \vec{AO} = \mathbf{b}$.
(iii) $\mathbf{a} + \mathbf{b} + \mathbf{c} = \vec{AB} + \vec{BC} + \vec{CD} = \vec{AC} + \vec{CD} = \vec{AD} = 2\vec{AO} = 2\mathbf{b}$.

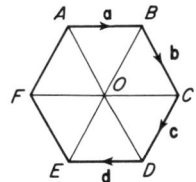

Fig. 4 Hexagon *ABCDEF*. 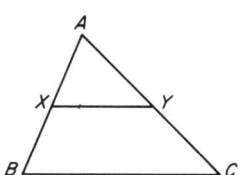 **Fig. 5** Triangle *ABC*.

*Modulus of a vector

The **modulus** or magnitude of a vector $\vec{AB}$ is its length. If $\vec{AB} = \mathbf{a}$, the modulus is written $|\vec{AB}|$, or $|\mathbf{a}|$.
A **unit** vector has length 1, so **a** is a unit vector if $|\mathbf{a}| = 1$.
The **zero** vector **0** has zero modulus, for example $\vec{AA} = \mathbf{0}$.
Parallel vectors **a** and **b** are multiples of each other, so $\mathbf{a} = k\mathbf{b}$, for some number k. Then $|\mathbf{a}| = k|\mathbf{b}|$, and the ratio of their lengths is $1:k$.

*Example In triangle *ABC*, *X* and *Y* are the midpoints of *AB* and *AC* respectively. Show that $\vec{XY}$ is parallel to $\vec{BC}$ and half its length.

ABC is drawn in Figure 5. $2\vec{XA} = \vec{BA}$ and $2\vec{AY} = \vec{AC}$, since *X* and *Y* are midpoints. Therefore, $\vec{BC} = \vec{BA} + \vec{AC} = 2\vec{XA} + 2\vec{AY} = 2(\vec{XA} + \vec{AY}) = 2\vec{XY}$.
So *BC* is parallel to *XY* and twice its length.

SAMPLE QUESTIONS
1 *ABCDEFGH* is a rectangular octagon, centre *O* (Figure 6). $\vec{AB} = \mathbf{w}$, $\vec{BC} = \mathbf{x}$, $\vec{CD} = \mathbf{y}$ and $\vec{DE} = \mathbf{z}$. Write down the following vectors in terms of some or all of **w**, **x**, **y** and **z**: (*a*) $\vec{EF}$, (*b*) $\vec{DF}$, (*c*) $\vec{BF}$, (*d*) $\vec{AO}$.
2 *PQRS* is a quadrilateral, see Figure 7, $\vec{PQ} = \mathbf{a}$, $\vec{QR} = \mathbf{b}$, $\vec{RS} = \mathbf{c}$, $\vec{SP} = \mathbf{d}$, $\vec{PR} = \mathbf{e}$ and $\vec{QS} = \mathbf{f}$. Write as a single vector: (*a*) $\mathbf{a} + \mathbf{b}$, (*b*) $\mathbf{e} - \mathbf{b}$, (*c*) $\mathbf{d} + \mathbf{f}$, (*d*) $\mathbf{a} + \mathbf{b} + \mathbf{c} - \mathbf{e}$.

159 *Vectors*

*3 In triangle ABC, X is the midpoint of BC. Show that $\overrightarrow{AB} + \overrightarrow{AC} = 2\overrightarrow{AX}$.

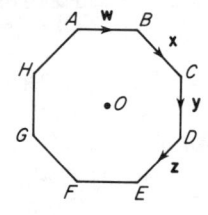

Fig. 6 Regular Octagon $ABCDEFGH$. 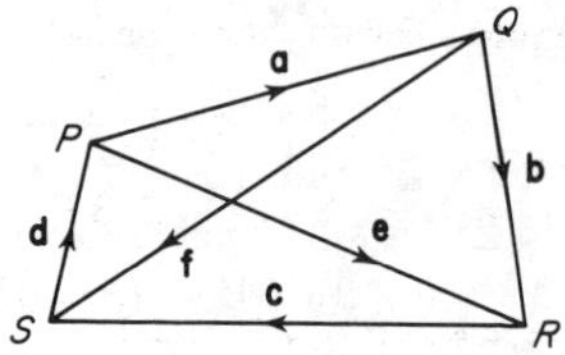   **Fig. 7** Quadrilateral $PQRS$.

*4 Show that the diagonals AC and BD of the parallelogram $ABCD$ bisect one another.

*5 In triangle OAB, $\overrightarrow{OA} = \mathbf{a}$, $\overrightarrow{OB} = \mathbf{b}$, X is the midpoint of OA and G is the point such that $3\overrightarrow{OG} = \mathbf{a} + \mathbf{b}$. In terms of $\mathbf{a}$ and $\mathbf{b}$, write down the vectors $\overrightarrow{BX}$ and $\overrightarrow{BG}$. Hence show that G lies on BX and find the ratio in which G divides BX.

Column vectors

Vectors in the Cartesian plane can be represented by a pair of numbers. These are the x and y displacements from the start to the end of the vector, say a and b respectively. The vector is written as a **column** $\begin{pmatrix} a \\ b \end{pmatrix}$, the two **components** represent a **across** and b **up**.

*When the vector is drawn on a unit grid, the **gradient** of the vector is b/a, and its **modulus** (length) is $\sqrt{(a^2 + b^2)}$.

*Example For each of the vectors in the diagram, Figure 8, write down: (*a*) the corresponding column vector, (*b*) the gradient, (*c*) the modulus, of the vector.

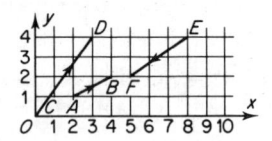

Fig. 8 Column vectors. 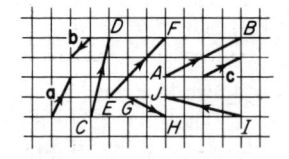   **Fig. 9** Column vectors.

Reading the x and y displacements from Figure 8,

(*a*) The column vectors are $\overrightarrow{AB} = \begin{pmatrix} 2 \\ 1 \end{pmatrix}$, $\overrightarrow{CD} = \begin{pmatrix} 3 \\ 4 \end{pmatrix}$, $\overrightarrow{EF} = \begin{pmatrix} -3 \\ -2 \end{pmatrix}$

(*b*) The gradients are $1/2$ $4/3$ $(-2)/(-3) = 2/3$

(*c*) The moduli are $\sqrt{(2^2 + 1^2)} = \sqrt{5}$ $\sqrt{(3^2 + 4^2)} = 5$
$\sqrt{(3^2 + 2^2)} = \sqrt{13}$

Column vectors represent displacements so they can be **added**, **subtracted**

Vectors 160

and **multiplied by a scalar** by operating on **each** of the components separately.

Example Given the vectors

$$\mathbf{a} = \begin{pmatrix} 2 \\ 4 \end{pmatrix} \quad \mathbf{b} = \begin{pmatrix} 3 \\ 5 \end{pmatrix} \quad \mathbf{c} = \begin{pmatrix} 7 \\ 1 \end{pmatrix} \quad \mathbf{d} = \begin{pmatrix} -3 \\ 2 \end{pmatrix} \quad \text{find:}$$

(a) $\mathbf{a} + \mathbf{b}$ (b) $\mathbf{b} - \mathbf{d}$ (c) $3\mathbf{a}$ (d) $2\mathbf{c} + \mathbf{d}$

(a) $\mathbf{a} + \mathbf{b} = \begin{pmatrix} 2 \\ 4 \end{pmatrix} + \begin{pmatrix} 3 \\ 5 \end{pmatrix} = \begin{pmatrix} 2+3 \\ 4+5 \end{pmatrix} = \begin{pmatrix} 5 \\ 9 \end{pmatrix}$

(b) $\mathbf{b} - \mathbf{d} = \begin{pmatrix} 3 \\ 5 \end{pmatrix} - \begin{pmatrix} -3 \\ 2 \end{pmatrix} = \begin{pmatrix} 3-(-3) \\ 5-2 \end{pmatrix} = \begin{pmatrix} 6 \\ 3 \end{pmatrix}$

(c) $3\mathbf{a} = 3\begin{pmatrix} 2 \\ 4 \end{pmatrix} = \begin{pmatrix} 3 \times 2 \\ 3 \times 4 \end{pmatrix} = \begin{pmatrix} 6 \\ 12 \end{pmatrix}$

(d) $2\mathbf{c} + \mathbf{d} = 2\begin{pmatrix} 7 \\ 1 \end{pmatrix} + \begin{pmatrix} -3 \\ 2 \end{pmatrix} = \begin{pmatrix} 2 \times 7 - 3 \\ 2 \times 1 + 2 \end{pmatrix} = \begin{pmatrix} 11 \\ 4 \end{pmatrix}$

SAMPLE QUESTIONS

6 Using Figure 9, (a) Write down $\overrightarrow{AB}, \overrightarrow{CD}, \overrightarrow{EF}, \overrightarrow{GH}, \overrightarrow{IJ}$ as column vectors.
 (b) Write as column vectors **a, b, c**.
 (c) List any pairs of parallel vectors amongst those above.
 *(d) List any sets of vectors with the same modulus.

7 On a grid draw the vectors, labelling the start and finish of each:

$$\overrightarrow{PQ} = \begin{pmatrix} 3 \\ 1 \end{pmatrix} \quad \overrightarrow{RS} = \begin{pmatrix} 2 \\ 4 \end{pmatrix} \quad \overrightarrow{TV} = \begin{pmatrix} 2 \\ -3 \end{pmatrix} \quad \overrightarrow{XY} = \begin{pmatrix} -1 \\ 4 \end{pmatrix}$$

$$\overrightarrow{AB} = \begin{pmatrix} 0 \\ 0 \end{pmatrix} \quad \overrightarrow{CD} = \begin{pmatrix} -2 \\ -3 \end{pmatrix}$$

8 Write down the column vectors with the given start and end points:
 (a) start (3, 7), end (6, 9),   (b) start (2, 1), end (8, 8),
 (c) start (7, 0), end (7, 11),   (d) start (1, 1), end (−3, 3),
 (e) start (6, 2), end (2, 6),   (f) start (−5, −2), end (3, −4).

9 $\mathbf{a} = \begin{pmatrix} 4 \\ -5 \end{pmatrix} \quad \mathbf{b} = \begin{pmatrix} 3 \\ 6 \end{pmatrix} \quad \mathbf{c} = \begin{pmatrix} -2 \\ 10 \end{pmatrix} \quad \mathbf{d} = \begin{pmatrix} 7 \\ 8 \end{pmatrix}$ Write down (a) $\mathbf{a} + \mathbf{b}$
 (b) $5\mathbf{c}$ (c) $\mathbf{b} - \mathbf{d}$ (d) $3\mathbf{c} + 4\mathbf{d}$ (e) $2\mathbf{b} - 3\mathbf{a}$ (f) $\mathbf{b} - 2\mathbf{c} + 3\mathbf{d}$
 *(g) $|\mathbf{a}|$

*10 Points P and Q have coordinates (4, 2) and (7, −2) relative to the origin O (0, 0). (a) Express $\overrightarrow{PQ}$ as a column vector and find its length. (b) Find the coordinates of the point R such that $\overrightarrow{RQ} = 2\overrightarrow{OP}$.

*11 On squared paper draw axes Oxy with scales for x from −3 to 7 and for y from −3 to 6. Plot the points A (−2, 3) and C (4, 1) on your diagram.
 (a) Find $\overrightarrow{AC}$ and $\overrightarrow{OB}$, given that $\overrightarrow{OB} = \overrightarrow{OA} + \overrightarrow{OC}$. (b) Plot the point B on your diagram and complete the figure $OABC$. (c) Find $\overrightarrow{OD}$ where D is the fourth vertex of parallelogram $OACD$.

VECTORS AND MATRICES
10.2 Matrices

10

Mary goes out to buy some fruit and makes a list of what she wants to buy: 4 apples, 6 oranges, 5 bananas. Tom also goes shopping to buy: 8 apples and 3 oranges. This information can be shown in a table, or more briefly, in a rectangular array of numbers, a **matrix**.

	Apples	Oranges	Bananas
Mary	4	6	5
Tom	8	3	0

$Matrix$
$$\begin{pmatrix} 4 & 6 & 5 \\ 8 & 3 & 0 \end{pmatrix}$$

This matrix consists of two **rows** and three **columns**, so has **order**, or size, 2×3, (in words two by three). The matrix of numbers is enclosed in large brackets whilst the row and column labels are written outside when they are needed.

An $n \times m$ matrix is a rectangular array of nm numbers arranged in n rows and m columns with order $n \times m$.

Addition and scalar multiplication

Suppose Mary and Tom's purchases of fruit the following week were written as the matrix $\begin{pmatrix} 3 & 0 & 5 \\ 6 & 3 & 0 \end{pmatrix}$. Then their total purchases over the two weeks are:

$\begin{pmatrix} 4 & 6 & 5 \\ 8 & 3 & 0 \end{pmatrix} + \begin{pmatrix} 3 & 0 & 5 \\ 6 & 3 & 0 \end{pmatrix} = \begin{pmatrix} 7 & 6 & 10 \\ 14 & 6 & 0 \end{pmatrix}$, showing the **addition** of matrices.

Suppose, on the other hand, that Mary and Tom bought the same numbers of apples, oranges and bananas each week for three weeks.
Then their total purchases over the three weeks are:

$3 \times \begin{pmatrix} 4 & 6 & 5 \\ 8 & 3 & 0 \end{pmatrix} = \begin{pmatrix} 12 & 18 & 15 \\ 24 & 9 & 0 \end{pmatrix}$, showing **scalar multiplication** of a matrix.

The arithmetic of matrices, addition, subtraction and scalar multiplication, is done **placewise**, that is, the arithmetic is done in **each row-column position** of the matrices. For example, it makes sense to add 'apples bought by Tom last week' to 'apples bought by Tom this week', but it would be pointless to add this to 'oranges bought by Mary this week'. So such arithmetic can only be carried out on matrices of the **same** order.

Matrices are given capital letters as names and are printed in heavy type in order to show they are rectangular arrays of numbers. This is the same notation as vectors, since vectors are simply matrices with one column.

Example Given the matrices.

$\mathbf{A} = \begin{pmatrix} 4 & 1 \\ 9 & 2 \end{pmatrix}$ $\mathbf{B} = \begin{pmatrix} 6 & 2 \\ 3 & 1 \end{pmatrix}$ $\mathbf{C} = \begin{pmatrix} 3 & 4 & 5 \\ 4 & 5 & 6 \end{pmatrix}$, find where possible:

(a) $2\mathbf{C}$ (b) $\mathbf{A} + \mathbf{B}$ (c) $\mathbf{A} + \mathbf{C}$ (d) $\mathbf{A} - \mathbf{B}$ (e) $2\mathbf{A} + 3\mathbf{B}$

(a) $2\mathbf{C} = 2 \times \begin{pmatrix} 3 & 4 & 5 \\ 4 & 5 & 6 \end{pmatrix} = \begin{pmatrix} 6 & 8 & 10 \\ 8 & 10 & 12 \end{pmatrix}$

(b) $\mathbf{A} + \mathbf{B} = \begin{pmatrix} 4 & 1 \\ 9 & 2 \end{pmatrix} + \begin{pmatrix} 6 & 2 \\ 3 & 1 \end{pmatrix} = \begin{pmatrix} 10 & 3 \\ 12 & 3 \end{pmatrix}$

(c) $\mathbf{A}$ is 2×2 and $\mathbf{C}$ is 2×3 so $\mathbf{A} + \mathbf{C}$ has no meaning.

(d) $\mathbf{A} - \mathbf{B} = \begin{pmatrix} 4 & 1 \\ 9 & 2 \end{pmatrix} - \begin{pmatrix} 6 & 2 \\ 3 & 1 \end{pmatrix} = \begin{pmatrix} -2 & -1 \\ 6 & 1 \end{pmatrix}$

(e) $2\mathbf{A} + 3\mathbf{B} = \begin{pmatrix} 2 \times 4 + 3 \times 6 & 2 \times 1 + 3 \times 2 \\ 2 \times 9 + 3 \times 3 & 2 \times 2 + 3 \times 1 \end{pmatrix} = \begin{pmatrix} 26 & 8 \\ 27 & 7 \end{pmatrix}$

Multiplication of matrices

Returning to the shopping list model, multiplication of a matrix by a matrix, can be defined as follows. Suppose that the prices of an apple, 9p, an orange, 12p, and a banana, 15p, are written as a 3×1 matrix:

$\begin{pmatrix} 9 \\ 12 \\ 15 \end{pmatrix}$ Then Mary's total bill for the first week can be found by multiplying her 1×3 fruit row (4 6 5), by this 3×1 price column matrix,

$(4 \ 6 \ 5) \begin{pmatrix} 9 \\ 12 \\ 15 \end{pmatrix} = (4 \times 9 + 6 \times 12 + 5 \times 15) = (183)$, giving a bill of £1.83.

Repeating for Tom's order, the multiplication of the 2×3 first week's fruit matrix with the 3×1 price matrix gives a 2×1 total bill matrix.

$\begin{pmatrix} 4 & 6 & 5 \\ 8 & 3 & 0 \end{pmatrix} \begin{pmatrix} 9 \\ 12 \\ 15 \end{pmatrix} = \begin{pmatrix} 4 \times 9 + 6 \times 12 + 5 \times 15 \\ 8 \times 9 + 3 \times 12 + 0 \times 15 \end{pmatrix} = \begin{pmatrix} 183 \\ 108 \end{pmatrix}$

Supposing they buy the same fruit the following week but the prices change to 10p for apples, 11p for oranges and 16p for bananas. Using a 3×2 price matrix where the second column gives the second week's prices, the product matrix will now give Mary and Tom's bill for each week.

$\begin{pmatrix} 4 & 6 & 5 \\ 8 & 3 & 0 \end{pmatrix} \begin{pmatrix} 9 & 10 \\ 12 & 11 \\ 15 & 16 \end{pmatrix} = \begin{pmatrix} 4 \times 9 + 6 \times 12 + 5 \times 15 & 4 \times 10 + 6 \times 11 + 5 \times 16 \\ 8 \times 9 + 3 \times 12 + 0 \times 15 & 8 \times 10 + 3 \times 11 + 0 \times 16 \end{pmatrix}$

$= \begin{pmatrix} 183 & 186 \\ 108 & 113 \end{pmatrix}$

*General definition

The row r, column c entry of the product **AB** of matrices **A** and **B**, is the **sum** of the **products of corresponding terms** in row r of **A** and column c of **B**. So the **length** of **A**'s **row** and **B**'s **column** must be equal.

163 *Matrices*

When **A** is an $m \times n$ matrix and **B** is a $p \times q$ matrix then the matrix product **AB** can be found if $n = p$, that is,

> the **number of columns** of **A** = the **number of rows** of **B**.

The orders of two matrices and their product fit the following pattern.

First Matrix **A**	Second Matrix **B**	Product **AB**
2×3	3×1	2×1
1×3	3×4	1×4
2×3	2×2	does not exist

*Example

$$\mathbf{A} = \begin{pmatrix} 2 & 1 & 2 \\ 0 & 2 & 0 \end{pmatrix} \quad \mathbf{B} = \begin{pmatrix} 2 \\ 0 \end{pmatrix} \quad \mathbf{C} = \begin{pmatrix} 1 & 3 \\ 0 & 2 \end{pmatrix} \quad \mathbf{D} = \begin{pmatrix} 2 & 1 \\ -1 & 0 \end{pmatrix}$$

Find if possible: (*a*) **CB**, (*b*) **BC**, (*c*) **CA**, (*d*) **CD**, (*e*) **DC**, (*f*) **(DC)A**.

(*a*) Checking the order of the product **CB**, 2×2 by 2×1 gives a 2×1 matrix:
$$\mathbf{CB} = \begin{pmatrix} 1 & 3 \\ 0 & 2 \end{pmatrix}\begin{pmatrix} 2 \\ 0 \end{pmatrix} = \begin{pmatrix} 1 \times 2 + 3 \times 0 \\ 0 \times 2 + 2 \times 0 \end{pmatrix} = \begin{pmatrix} 2 \\ 0 \end{pmatrix}$$

(*b*) The product **BC does not exist**, since **B** has 1 column and **C** has 2 rows.

(*c*) **CA**: 2×2 by 2×3 gives a 2×3 product matrix:
$$\mathbf{CA} = \begin{pmatrix} 1 & 3 \\ 0 & 2 \end{pmatrix}\begin{pmatrix} 2 & 1 & 2 \\ 0 & 2 & 0 \end{pmatrix} = \begin{pmatrix} 1 \times 2 + 3 \times 0 & 1 \times 1 + 3 \times 2 & 1 \times 2 + 3 \times 0 \\ 0 \times 2 + 2 \times 0 & 0 \times 1 + 2 \times 2 & 0 \times 2 + 2 \times 0 \end{pmatrix}$$
$$= \begin{pmatrix} 2 & 7 & 2 \\ 0 & 4 & 0 \end{pmatrix}$$

(*d*) and (*e*) **CD** and **DC**: 2×2 by 2×2 gives a 2×2 product matrix:
$$\mathbf{CD} = \begin{pmatrix} 1 & 3 \\ 0 & 2 \end{pmatrix}\begin{pmatrix} 2 & 1 \\ -1 & 0 \end{pmatrix} = \begin{pmatrix} 1 \times 2 + 3 \times (-1) & 1 \times 1 + 3 \times 0 \\ 0 \times 2 + 2 \times (-1) & 0 \times 1 + 2 \times 0 \end{pmatrix} = \begin{pmatrix} -1 & 1 \\ -2 & 0 \end{pmatrix}$$
$$\mathbf{DC} = \begin{pmatrix} 2 & 1 \\ -1 & 0 \end{pmatrix}\begin{pmatrix} 1 & 3 \\ 0 & 2 \end{pmatrix} = \begin{pmatrix} 2 \times 1 + 1 \times 0 & 2 \times 3 + 1 \times 2 \\ (-1) \times 1 + 0 \times 0 & (-1) \times 3 + 0 \times 2 \end{pmatrix}$$
$$= \begin{pmatrix} 2 & 8 \\ -1 & -3 \end{pmatrix}$$

(*f*) To find **(DC)A** multiply the answer to part (*e*) by **A** as follows:
$$\mathbf{(DC)A} = \begin{pmatrix} 2 & 8 \\ -1 & -3 \end{pmatrix}\begin{pmatrix} 2 & 1 & 2 \\ 0 & 2 & 0 \end{pmatrix}$$

Matrices 164

$$= \begin{pmatrix} 2\times2+8\times0 & 2\times1+8\times2 & 2\times2+8\times0 \\ -1\times2+(-3)\times0 & -1\times1+(-3)\times2 & -1\times2+(-3)\times0 \end{pmatrix}$$

$$= \begin{pmatrix} 4 & 18 & 4 \\ -2 & -7 & -2 \end{pmatrix}$$

SAMPLE QUESTIONS

1 Find the values of a, b, c and d so that
$$\begin{pmatrix} a & -1 \\ 6 & 2 \end{pmatrix} + \begin{pmatrix} 2 & b \\ c & -7 \end{pmatrix} = \begin{pmatrix} 5 & -5 \\ 3 & d \end{pmatrix}$$

2 $\mathbf{P} = \begin{pmatrix} 1 & 0 \\ 0 & -3 \\ 3 & 1 \end{pmatrix}$ and $\mathbf{Q} = \begin{pmatrix} 1 & -1 & 2 \\ 2 & -3 & 4 \end{pmatrix}$ Find (a) **PQ** (b) **QP**

3 $\mathbf{A} = \begin{pmatrix} 1 & 0 \\ 0 & 2 \end{pmatrix}$ $\mathbf{B} = \begin{pmatrix} -1 & 1 \\ 0 & -2 \end{pmatrix}$ $\mathbf{C} = \begin{pmatrix} 2 & -3 \\ -1 & 0 \end{pmatrix}$ Work out

(a) **AB** (b) **BA** (c) **BC** (d) **CB** (e) **(AB)C** (f) **A(BC)**

4 Find the values of p and q so that $\begin{pmatrix} p & 5 \\ 4 & q \end{pmatrix}\begin{pmatrix} 2 \\ 1 \end{pmatrix} = \begin{pmatrix} 7 \\ 3 \end{pmatrix}$

*5 The matrix **A** has order 2×3.
  (a) If $\mathbf{A} + \mathbf{B} = \mathbf{C}$ what can you say about the order of (i) **B**, (ii) **C**?
  (b) If $\mathbf{AD} = \mathbf{E}$ what can you say about the order of (i) **D**, (ii) **E**?

6 Faith, Kevin and Ian go shopping for fireworks. Faith buys 4 Roman candles, 3 bomb-bursts, and 8 sparklers. Kevin buys 2 Roman candles, 6 bomb-bursts and 12 bangers. Ian buys 20 bangers and 5 bomb-bursts. Represent their purchases in a 3×4 matrix, **S**. Roman candles cost 23p, bomb-bursts cost 31p, sparklers cost 3p and bangers cost 9p. Represent the prices of the four types of fireworks in a 4×1 column matrix, **P**, and calculate the matrix product **SP**.
How much did Faith spend on fireworks? Who spent the most? Multiply the row vector (1 1 1) by the matrix **SP** and interpret the result.

7 Three lorries A, B, C carry crates of beer, cider, lemonade and coke. Lorry A carries 60 crates of beer, 25 crates of cider and 20 crates of lemonade. Lorry B carries 70 crates of beer and 30 crates of coke. Lorry C carries 30 crates of cider, 10 crates of lemonade and 25 crates of coke.
(a) Express this information in a 3×4 matrix.
(b) A crate of beer weighs 12kg, a crate of cider weighs 10kg, and each crate of lemonade and coke weigh 9kg. Express these weights as a 4×1 matrix.
(c) Use your matrices to find the weight in kilograms carried by each lorry.

*8 $\mathbf{A} = \begin{pmatrix} 1 & -2 \\ 2 & 3 \end{pmatrix}$ $\mathbf{B} = \begin{pmatrix} -1 & 2 \\ -2 & -3 \end{pmatrix}$ Evaluate (a) $\mathbf{A} + \mathbf{B}$ (b) $\mathbf{AB}$ (c) $\mathbf{A}^2$.

VECTORS AND MATRICES
10.3 Route Matrices

Routes
When travelling between places there may be several **routes** from one place to another. Suppose there are 3 routes between X and Y, 2 between X and Z, 4 between Y and Z and 1 loop from X to X, as shown in Figure 1.

$$\text{from} \begin{array}{c} \\ X \\ Y \\ Z \end{array} \begin{pmatrix} \overset{X}{1} & \overset{to}{\overset{Y}{3}} & \overset{Z}{2} \\ 3 & 0 & 4 \\ 2 & 4 & 0 \end{pmatrix}$$

Fig. 1 (a) Road network.   (b) Route matrix, **M**.

This information may be written in a matrix, **M**, with rows and columns labelled XYZ. The entry in row X, column Y gives the number of routes from X to Y, in this case 3 routes, and all the routes are **two-way**, from X to Y and from Y to X. The matrix is therefore **symmetrical** about a diagonal line running top left to bottom right. If some of these routes were only **one-way** then the matrix would no longer be symmetrical.

Two-stage journeys
Using the network above, count how many routes there are from X to Y in two **stages**.

X to X 1 route X to Y 3 routes X to X to $Y = 1 \times 3 = 3$ routes.
X to Y 3 routes Y to Y 0 routes X to Y to $Y = 3 \times 0 = 0$ routes.
X to Z 2 routes Z to Y 4 routes X to Z to $Y = 2 \times 4 = 8$ routes.

So the total number of 2-stage routes from X to Y is $3 + 0 + 8 = 11$. This is the product of the X row with the Y column of the route matrix, **M**, so is the X, Y entry in the matrix $\mathbf{MM} = \mathbf{M}^2$.

The matrix $\mathbf{M}^2$ represents the **2-stage route matrix** for this network.

$$\mathbf{M}^2 = \mathbf{M} \times \mathbf{M} = \begin{pmatrix} 1 & 3 & 2 \\ 3 & 0 & 4 \\ 2 & 4 & 0 \end{pmatrix} \begin{pmatrix} 1 & 3 & 2 \\ 3 & 0 & 4 \\ 2 & 4 & 0 \end{pmatrix} = \begin{pmatrix} 14 & 11 & 14 \\ 11 & 25 & 6 \\ 14 & 6 & 20 \end{pmatrix}$$

one-stage   two-stage

The ideas of route matrices can be applied to other situations.
One example concerns the results of games between a group of players.

*Example Five men played ten matches of squash with the results:
Ahmed beat Calum, Donald and Eric. Ben beat Ahmed, Eric and Donald. Calum beat Donald and Ben, and Eric beat Calum and Donald.
Draw a network showing these dominances and compile its route matrix, **S**.
Calculate $\mathbf{S}^2$, find the total of the one and two-stage dominances for each player, and hence rank them in order.

Figure 2 shows the dominance network, an arrow from Ahmed to Calum means Ahmed beat Calum, and the one and two-stage route matrices for this network.

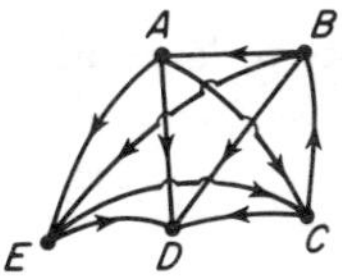

Route Matrices   166

$$\mathbf{S} \begin{array}{c} \\ A \\ B \\ C \\ D \\ E \end{array} \begin{pmatrix} A & B & C & D & E \\ 0 & 0 & 1 & 1 & 1 \\ 1 & 0 & 0 & 1 & 1 \\ 0 & 1 & 0 & 1 & 0 \\ 0 & 0 & 0 & 0 & 0 \\ 0 & 0 & 1 & 1 & 0 \end{pmatrix} \quad \mathbf{S}^2 \begin{array}{c} \\ A \\ B \\ C \\ D \\ E \end{array} \begin{pmatrix} A & B & C & D & E \\ 0 & 1 & 1 & 2 & 0 \\ 0 & 0 & 2 & 2 & 1 \\ 1 & 0 & 0 & 1 & 1 \\ 0 & 0 & 0 & 0 & 0 \\ 0 & 1 & 0 & 1 & 0 \end{pmatrix}$$

Fig. 2 Dominance networks and matrix. 10 squash games between 5 players.

Adding the rows of **S** and $\mathbf{S}^2$ give the number of one and two stage dominances for each player:

Ahmed $3 + 4 = 7$, Ben $3 + 5 = 8$, Calum $2 + 3 = 5$,
Donald $0 + 0 = 0$, Eric $2 + 2 = 4$.

The rank order should therefore be **Ben, Ahmed, Calum, Eric, Donald**.

SAMPLE QUESTIONS

1 Figure 3 shows the routes between P, Q and R. Write down the route matrix **M** and calculate the 2-stage route matrix $\mathbf{M}^2$. Hence write down the number of 2-stage routes from P to R.

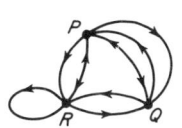

$$\mathbf{X} \begin{array}{c} \\ L \\ M \\ B \\ S \end{array} \begin{pmatrix} L & M & B & S \\ 0 & 2 & 3 & 2 \\ 3 & 0 & 4 & 0 \\ 0 & 1 & 0 & 0 \\ 1 & 2 & 1 & 0 \end{pmatrix}$$

Fig. 3 Road network between P, Q, R.   **Fig. 4** Flight paths matrix **X**.

2 The matrix **X**, given in Figure 4, represents a network of flight paths between London, Manchester, Birmingham and Shannon airports. (*a*) Draw the network of flight paths. (*b*) Is it possible to reach Shannon from Birmingham? If so, describe the route, if not, give a reason.

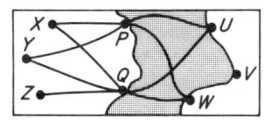

Fig. 5 Road and ferry links. 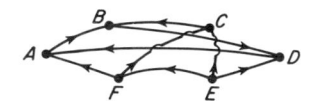   **Fig. 6** Tennis tournament.

3 Figure 5 shows the road links from towns X, Y, Z to ports P and Q, and the ferry routes from P, Q to ports U, V and W. Compile route matrices, (*a*) **R** from X, Y, Z to P, Q, (*b*) **F** from P, Q to U, V, W (*c*) **RF** from X, Y, Z to U, V, W. How many routes are there from X to W?

*4 Figure 6 shows the wins between 6 girls, Alison, Beth, Cath, Dotty, Enid and Fanny, who played 3 matches each. Calculate the one- and two-stage dominance matrices and use the row totals to rank the girls in order.

VECTORS AND MATRICES
10.4 *Matrix transformations

10

The position of a point $P(4, 1)$, in the Cartesian plane is given by its **position vector** $\begin{pmatrix} 4 \\ 1 \end{pmatrix}$ the vector from the origin to P.

Multiplying this vector by a 2×2 matrix $\mathbf{M} \begin{pmatrix} 1 & 2 \\ -1 & 3 \end{pmatrix}$ transforms this position vector $\overrightarrow{OP}$ to another position vector $\overrightarrow{OP'}$, so P goes to P'.

$$\mathbf{M} \times \overrightarrow{OP} = \overrightarrow{OP'} \qquad \begin{pmatrix} 1 & 2 \\ -1 & 3 \end{pmatrix}\begin{pmatrix} 4 \\ 1 \end{pmatrix} = \begin{pmatrix} 6 \\ -1 \end{pmatrix} \quad P(4, 1) \text{ goes to } P'(6, -1)$$

*Example The rectangle $ABCD$, $A(1, 2)$, $B(3, 2)$, $C(3, 3)$, $D(1, 3)$, is transformed to $A'B'C'D'$ by the matrix $\mathbf{M}$, where $\mathbf{M} = \begin{pmatrix} 0 & -1 \\ 1 & 0 \end{pmatrix}$ Draw a graph showing $ABCD$ and its image $A'B'C'D'$. Describe the transformation $\mathbf{M}$.

Form a 2×4 matrix containing the four position vectors $\overrightarrow{OA}, \overrightarrow{OB}, \overrightarrow{OC}, \overrightarrow{OD}$, where O is the origin $(0, 0)$. Multiplying this matrix by $\mathbf{M}$, gives:

$$\begin{array}{cccc} \mathbf{M} & A \ B \ C \ D & & A' \ B' \ C' \ D' \end{array}$$
$$\begin{pmatrix} 0 & -1 \\ 1 & 0 \end{pmatrix} \begin{pmatrix} 1 & 3 & 3 & 1 \\ 2 & 2 & 3 & 3 \end{pmatrix} = \begin{pmatrix} -2 & -2 & -3 & -3 \\ 1 & 3 & 3 & 1 \end{pmatrix}$$

so $A(1, 2)$ goes to $A'(-2, 1)$.

Draw a set of axes, plot and join $ABCD$ and $A'B'C'D'$, see Figure 1. From the diagram transformation $\mathbf{M}$ is a **rotation about $(0, 0)$ through $90°$** (anticlockwise).

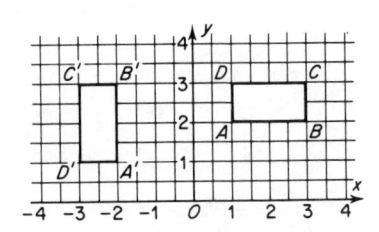

Fig. 1 Transformation $\mathbf{M}$ on $ABCD$.

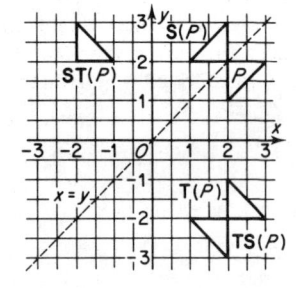

Fig. 2 Flag P under $\mathbf{S}$ and $\mathbf{T}$.

Two transformations of a shape may be done, one after the other, and the matrices may then be multiplied to find the matrix for the single combined transformation, as demonstrated in the next example.

Matrix Transformations

*Example Flag P is formed from the points $(2, 0)$, $(2, 2)$, $(3, 2)$ and $(2, 1)$. Transformation matrices **S** and **T** are defined as follows:

$$\mathbf{S} = \begin{pmatrix} 0 & 1 \\ 1 & 0 \end{pmatrix} \quad \mathbf{T} = \begin{pmatrix} 1 & 0 \\ 0 & -1 \end{pmatrix} \quad \text{Draw flag } P \text{ on axes from } -3 \text{ to } 3 \text{ for } x \text{ and } y.$$

Calculate the positions of, (a) $\mathbf{S}(P)$, (b) $\mathbf{T}(P)$, (c) $\mathbf{TS}(P)$, (d) $\mathbf{ST}(P)$. Draw and label these images on the axes. Describe the transformations **S**, **T**, **TS** and **ST**. Give a single transformation which maps $\mathbf{ST}(P)$ to $\mathbf{TS}(P)$.

Figure 2 shows the axes and flag P drawn. The calculation of the positions of the images of P under S and T is shown below.

$$\begin{array}{ccc} \mathbf{S} & P & \mathbf{S}(P) \\ \begin{pmatrix} 0 & 1 \\ 1 & 0 \end{pmatrix} & \begin{pmatrix} 2 & 2 & 3 & 2 \\ 0 & 2 & 2 & 1 \end{pmatrix} = & \begin{pmatrix} 0 & 2 & 2 & 1 \\ 2 & 2 & 3 & 2 \end{pmatrix} \end{array}$$

$$\begin{array}{ccc} \mathbf{T} & P & \mathbf{T}(P) \\ \begin{pmatrix} 1 & 0 \\ 0 & -1 \end{pmatrix} & \begin{pmatrix} 2 & 2 & 3 & 2 \\ 0 & 2 & 2 & 1 \end{pmatrix} = & \begin{pmatrix} 2 & 2 & 3 & 2 \\ 0 & -2 & -2 & -1 \end{pmatrix} \end{array}$$

$$\begin{array}{ccc} \mathbf{T} & \mathbf{S}(P) & \mathbf{TS}(P) \\ \begin{pmatrix} 1 & 0 \\ 0 & -1 \end{pmatrix} & \begin{pmatrix} 0 & 2 & 2 & 1 \\ 2 & 2 & 3 & 2 \end{pmatrix} = & \begin{pmatrix} 0 & 2 & 2 & 1 \\ -2 & -2 & -3 & -2 \end{pmatrix} \end{array}$$

$$\begin{array}{ccc} \mathbf{S} & \mathbf{T}(P) & \mathbf{ST}(P) \\ \begin{pmatrix} 0 & 1 \\ 1 & 0 \end{pmatrix} & \begin{pmatrix} 2 & 2 & 3 & 2 \\ 0 & -2 & -2 & -1 \end{pmatrix} = & \begin{pmatrix} 0 & -2 & -2 & -1 \\ 2 & 2 & 3 & 2 \end{pmatrix} \end{array}$$

Once these images are drawn on the graph it is clear that, **S** is a **reflection in** $x = y$, **T** is a **reflection in the x-axis**, **TS** is a **rotation of** $-90°$ **about** $(0, 0)$ and **ST** is a **rotation of** $+90°$ **about** $(0, 0)$. A rotation of $180°$ about $(0, 0)$ will transform $\mathbf{ST}(P)$ to $\mathbf{TS}(P)$.

*Unit vector method

There is a quicker way of finding the transformed points which also leads to a way of finding the matrix of a specified transformation.

Consider the **unit square** $OACB$ with vertices at $O(0, 0)$, $A(1, 0)$, $C(1, 1)$, $B(0, 1)$ transformed by the matrix **M**, where

$$\mathbf{M} = \begin{pmatrix} a & b \\ c & d \end{pmatrix}. \quad \text{Then } \begin{array}{c}\mathbf{M}\\\begin{pmatrix} a & b \\ c & d \end{pmatrix}\end{array} \begin{array}{c}A\\\begin{pmatrix} 1 \\ 0 \end{pmatrix}\end{array} = \begin{array}{c}\mathbf{M}(A)\\\begin{pmatrix} a \\ c \end{pmatrix}\end{array} \text{ and } \begin{array}{c}\mathbf{M}\\\begin{pmatrix} a & b \\ c & d \end{pmatrix}\end{array} \begin{array}{c}B\\\begin{pmatrix} 0 \\ 1 \end{pmatrix}\end{array} = \begin{array}{c}\mathbf{M}(B)\\\begin{pmatrix} b \\ d \end{pmatrix}\end{array}$$

So the images of the unit vectors $\overrightarrow{OA}\begin{pmatrix}1\\0\end{pmatrix}$ and $\overrightarrow{OB}\begin{pmatrix}0\\1\end{pmatrix}$ may be read from the columns $\begin{pmatrix}a\\c\end{pmatrix}$ and $\begin{pmatrix}b\\d\end{pmatrix}$ of the transformation matrix M.

*Example Describe the transformation given by **M**, where $\mathbf{M} = \begin{pmatrix} 0 & -1 \\ 1 & 0 \end{pmatrix}$. Find the matrix of **R**, a rotation $270°$ about $(0, 0)$.

*Matrix Transformations

The transformation sends the unit vectors $\begin{pmatrix}1\\0\end{pmatrix}\begin{pmatrix}0\\1\end{pmatrix}$ to $\begin{pmatrix}0\\-1\end{pmatrix}\begin{pmatrix}-1\\0\end{pmatrix}$ reading from the two columns of the matrix **M**. Figure 3(a) shows that the transformation is a **reflection in the line** $x+y=0$.

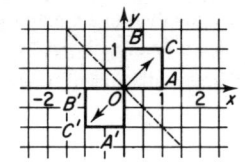

Fig. 3 (a) Transformation **M**.   (b) Transformation **R**.

Figure 3(b) shows the unit square $OACB$ and its image after transformation **R**. The unit vectors $\overrightarrow{OA}$, $\overrightarrow{OB}$ go to $\overrightarrow{OA'}$, $\overrightarrow{OB'}$.

The images of $\begin{pmatrix}1\\0\end{pmatrix}\begin{pmatrix}0\\1\end{pmatrix}$ are $\begin{pmatrix}0\\-1\end{pmatrix}\begin{pmatrix}1\\0\end{pmatrix}$ so the matrix of **R** is $\begin{pmatrix}0 & 1\\-1 & 0\end{pmatrix}$

*Inverse matrices

When a transformation T has a matrix **M** then the inverse transformation T^{-1} has matrix $\mathbf{M}^{-1}$, the **inverse matrix** of **M**.

If $\mathbf{M} = \begin{pmatrix}a & b\\c & d\end{pmatrix}$ then $\mathbf{M}^{-1} = \frac{1}{D}\begin{pmatrix}d & -b\\-c & a\end{pmatrix}$ where $D = ad - bc$ is the determinant of the matrix.

If **M** is a general 2×2 matrix and $D = 0$ then **M** has **no inverse** matrix.

*Example Find $\mathbf{M}^{-1}$, where $\mathbf{M} =$ (a) $\begin{pmatrix}-1 & 0\\0 & 1\end{pmatrix}$ (b) $\begin{pmatrix}1 & 3\\2 & 8\end{pmatrix}$.

(a) Since **M** is the matrix of a reflection in the y-axis, the inverse $\mathbf{M}^{-1}$ must be the same as **M**. Check $\begin{pmatrix}-1 & 0\\0 & 1\end{pmatrix}\begin{pmatrix}-1 & 0\\0 & 1\end{pmatrix} = \begin{pmatrix}1 & 0\\0 & 1\end{pmatrix}$ the identity matrix.

(b) The determinant $D = 1 \times 8 - 3 \times 2 = 2$, so
$$\mathbf{M}^{-1} = \frac{1}{2}\begin{pmatrix}8 & -3\\-2 & 1\end{pmatrix} = \begin{pmatrix}4 & -\frac{3}{2}\\-1 & \frac{1}{2}\end{pmatrix}$$

Combined transformations

*Example On a Cartesian graph, draw and label the triangle T, with vertices $O(0,0)$, $A(1,0)$, $B(1,2)$ and its image, $OA'B'$ after the transformation by the matrix **M**, where $\mathbf{M} = \begin{pmatrix}0 & -3\\3 & 0\end{pmatrix}$ Find and describe an enlargement **E** and a rotation **R** so that **RE** also transforms OAB to $OA'B'$. Write down the matrices of the transformations **E** and **R**, and verify that $\mathbf{M} = \mathbf{RE}$.

Matrix Transformations 170

Multiplying **M** by the position vectors of O, A and B gives the position vectors of O', A' and B'.

$$\begin{pmatrix} 0 & -3 \\ 3 & 0 \end{pmatrix} \begin{pmatrix} 0 & 1 & 1 \\ 0 & 0 & 2 \end{pmatrix} = \begin{pmatrix} 0 & 0 & -6 \\ 0 & 3 & 3 \end{pmatrix}$$ so $A' = (0, 3)$, $B' = (-6, 3)$, see Figure 4.

$OA'B'$ is similar to OAB, but enlarged by a factor $OA'/OA = 3/1 = 3$. The **enlargement, E, centre (0, 0), scale factor 3**, transforms OAB to $(0, 0)$, $(3, 0)$, $(3, 6)$, and a **rotation, R, about (0, 0), through $+90°$**, will transform this shape on to $OA'B'$.

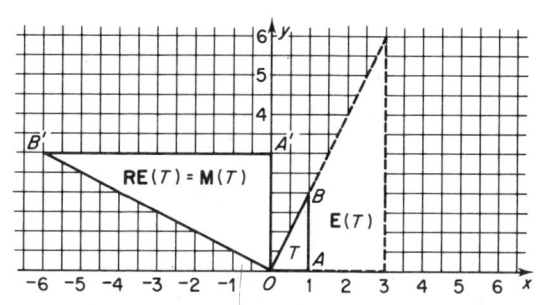

Fig. 4 **M = RE**.

Enlargement **E**, sends $\begin{pmatrix} 1 \\ 0 \end{pmatrix} \begin{pmatrix} 0 \\ 1 \end{pmatrix}$ to $\begin{pmatrix} 3 \\ 0 \end{pmatrix} \begin{pmatrix} 0 \\ 3 \end{pmatrix}$ so matrix $\mathbf{E} = \begin{pmatrix} 3 & 0 \\ 0 & 3 \end{pmatrix}$

Rotation **R**, sends $\begin{pmatrix} 1 \\ 0 \end{pmatrix} \begin{pmatrix} 0 \\ 1 \end{pmatrix}$ to $\begin{pmatrix} 0 \\ 1 \end{pmatrix} \begin{pmatrix} -1 \\ 0 \end{pmatrix}$ so matrix $\mathbf{R} = \begin{pmatrix} 0 & -1 \\ 1 & 0 \end{pmatrix}$

$\mathbf{RE} = \begin{pmatrix} 0 & -1 \\ 1 & 0 \end{pmatrix} \begin{pmatrix} 3 & 0 \\ 0 & 3 \end{pmatrix} = \begin{pmatrix} 0 & -3 \\ 3 & 0 \end{pmatrix} = \mathbf{M}$

SAMPLE QUESTIONS

*1 Draw the image of the unit square $OACB$, $O(0, 0)$, $A(1, 0)$, $B(0, 1)$, $C(1, 1)$ under the transformation with matrix:

(a) $\begin{pmatrix} 0 & -1 \\ 1 & 0 \end{pmatrix}$ (b) $\begin{pmatrix} 0 & 1 \\ -1 & 0 \end{pmatrix}$ (c) $\begin{pmatrix} 2 & 0 \\ 0 & 2 \end{pmatrix}$ (d) $\begin{pmatrix} 0 & -1 \\ -1 & 0 \end{pmatrix}$ (e) $\begin{pmatrix} -3 & 0 \\ 0 & -3 \end{pmatrix}$

*2 Describe the transformations drawn in 1 above.

*3 Use the unit vector method to find the matrix representing:
(a) reflection in the line $x = -y$, (b) rotation $180°$ about $(0, 0)$,
(c) enlargement, scale factor 5, centre $(0, 0)$ followed by a rotation $-90°$ about $(0, 0)$.

*4 Transformation **T** maps the point $(1, 0)$ to $(3, 4)$ and $(0, 1)$ to $(-4, 3)$. Find the matrix representing **T**. Find the image of $(3, 4)$ under the transformation **T**. Find the point which maps to $(4, -3)$ under **T**. **T** can be split into an enlargement followed by a rotation. Carefully describe these two transformations. Find the matrix representing $\mathbf{T}^{-1}$.

ANSWERS

Whole Numbers
1(a) {1, 2, 3, 4, 6, 8, 12, 24} (b) {1, 3, 5, 9, 15, 45} (c) {1, 2, 4, 7, 8, 14, 28, 56} (d) {1, 3, 5, 15, 25, 75} **2**(a) 60 (b) 147 (c) 125 **3**(a) $2^4 \times 3$ (b) $2^4 \times 5$ (c) $2^3 \times 3 \times 5$ **4**(a) LCM 175, HCF 5 (b) LCM 672, HCF 8 **5** 36 **6** 60 **7**(a) $\times 2$, 64 128 256 (b) -3, 20 17 14 (c) $+5$, 46 51 56 (d) Add last number, 47 76 123 (e) $\times 2$, 96 192 384 (f) add 2 more, 41 55 71 (g) $\times 11$, 161051 1771561 19487171 (h) -789, 6843 6054 5265 **8**(a) 4 (b) 3 (c) 63 (d) 5 **9**(a) {15, 12} (b) 1 (c) 4

Decimals
1(a) thousands (b) hundredths (c) ten millions (d) millionths **2** (a) 15.41 (b) 0.233244 (c) 0.89 (d) 6400 000 **3** 77 **4** £97.27, £2.73 **5** £990 000 **6** 19 **7** 586, £1752 **8** 23p **9** 105 000, £2.90 **10**(a) 1.732 (b) 4.472 (c) 9.950 (d) 0.8367 (e) 31.62 (f) 0.04472 **11**(a) 0.56 (b) 0.43 (c) 0.42 (d) 0.39 **12**(a) 5639 (b) 0.02208 (c) 25 740 000 (d) 0.000 2467 **13**(a) 3.82×10^{-3} (b) 2.49563×10^7 (c) $2.003\,284\,964 \times 10^4$ (d) 3.8984×10^{-5} **14**(a) 2 (b) -4 (c) 6 (d) 238 700 (e) 9.186 (f) 1.729 438 6 **15**(a) 6.8846×10^3 (b) $2.373\,9703 \times 10^4$ (c) 5.846×10^{-5} (d) 1.03×10^{-2} (e) 7.4088×10^{16} **16** 7.5×10^5 **17** 9.05 s (2DP)

Fractions
1(a) $\frac{4}{5}$ (b) $1\frac{2}{7}$ (c) $\frac{1}{4}$ (d) $\frac{13}{21}$ **2**(a) $\frac{5}{9}$, $\frac{4}{7}$ (b) $\frac{6}{5}$, $\frac{11}{9}$, $\frac{13}{10}$ (c) $\frac{1}{3}$, $\frac{7}{20}$, $\frac{12}{29}$, $\frac{4}{9}$ **3**(a) £38 (b) 2 kg (c) 378 **4**(a) $\frac{2}{5}$ (b) $\frac{1}{7}$ (c) $\frac{4}{23}$ (d) $\frac{2}{41}$ **5**(a) $1\frac{17}{45}$ (b) $\frac{3}{20}$ (c) $\frac{5}{16}$ (d) $5\frac{3}{4}$ (e) $\frac{7}{25}$ (f) $2\frac{1}{10}$ (g) $\frac{3}{20}$ (h) $\frac{3}{8}$ (i) $\frac{2}{29}$ **6**(a) 0.833 (b) 0.778 (c) 0.480 (d) 0.857 (e) 3.875 (f) 3.143 **7** Magazines £0.80, sweets £1.20, left £0.40 **8** Planes 14, helicopters 21, $\frac{5}{13}$ **9** George £250, Henry £200, Jean £200, Penelope £350, $\frac{7}{20}$ **10** 32.0625 g, 35.269 g, $\frac{19}{80}$. **11**(a) £4200 (b) £3360 (c) £2150.40, $\frac{48}{125}$.

Sets of Numbers
1 $\sqrt{2.5}, 2+\sqrt{3}$ **3**(a) -3 (b) 3 (c) $\sqrt{2}$ **4**(a) T (b) T (c) F (d) F (e) T

Directed Numbers
1 $-6°C$ **2** $-£23.43$, £1.57 **3**(a) -12 (b) -10 (c) -343 (d) $5\frac{4}{5}$ (e) 96 **5**(a) 5 (b) -7 **6**(a) -25 (b) -14 **7** (a) $\begin{pmatrix} -1 \\ -1 \\ -16 \end{pmatrix}$ (b) $\begin{pmatrix} 2 \\ 13 \\ -18 \end{pmatrix}$ **8** (a) $-2, -2, 0, 4, 10, 18$ (b) $-11, 1, 5, -11, -31$ (c) $-4, 4, 0, -4, 4, 36$ **9** $(4, -6), (-1, 2), (-11, 15)$

Powers and Roots
1(a) 243 (b) 343 (c) 1 000 000 (d) 1 **2**(a) 2^6 (b) 3^4 (c) 8^2 (d) 10^5 (e) 2^{10} (f) 5^4 **3**(a) 2.65 (b) 3.16 (c) 2.92 (d) 1.26 (e) 31.6 (f) 10.1 (g) 7.07 **4**(a) 2.24 (b) 4.47 (c) 2.2 **5**(a) 2^{12} (b) 3^3 (c) 4^{10} (d) 15^2 (e) x^2 (f) $3y^5$ **6**(a) 1.92 (b) 3.92×10^{13} (c) 4×10^8 (d) 8×10^{-11} **7**(a) 9 (b) $\frac{1}{10}$ (c) 81 (d) $\frac{1}{343}$ (e) $\frac{1}{2}$ (f) 243 (g) 128 (h) $\frac{1}{100\,000}$ (i) $\frac{1}{32}$

Answers 172

Know your Calculator

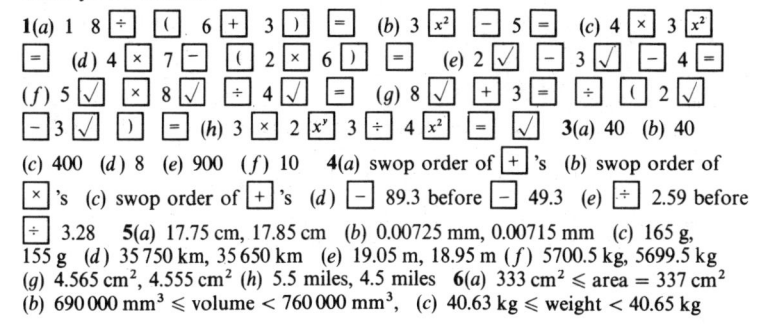

1(a) 1 8 ÷ (6 + 3) =   (b) 3 x^2 − 5 =   (c) 4 × 3 x^2
 =   (d) 4 × 7 − (2 × 6) =   (e) 2 $\sqrt{}$ − 3 $\sqrt{}$ − 4 =
(f) 5 $\sqrt{}$ × 8 $\sqrt{}$ ÷ 4 $\sqrt{}$ =   (g) 8 $\sqrt{}$ + 3 = ÷ (2 $\sqrt{}$
 − 3 $\sqrt{}$) =   (h) 3 × 2 x^y 3 ÷ 4 x^2 = $\sqrt{}$ 3(a) 40 (b) 40
(c) 400 (d) 8 (e) 900 (f) 10   4(a) swop order of + 's (b) swop order of
 × 's (c) swop order of + 's (d) − 89.3 before − 49.3 (e) ÷ 2.59 before
 ÷ 3.28   5(a) 17.75 cm, 17.85 cm (b) 0.00725 mm, 0.00715 mm (c) 165 g,
155 g (d) 35 750 km, 35 650 km (e) 19.05 m, 18.95 m (f) 5700.5 kg, 5699.5 kg
(g) 4.565 cm², 4.555 cm² (h) 5.5 miles, 4.5 miles 6(a) 333 cm² ⩽ area = 337 cm²
(b) 690 000 mm³ ⩽ volume < 760 000 mm³, (c) 40.63 kg ⩽ weight < 40.65 kg

Ratio and Proportion

1(a) 5:4 (b) 1:12 (c) 1:14 (d) 1:14   2(a) 4000 (b) 16 (c) 8640 (d) 2000
(e) 3.4 (f) 937.5   3(a) 750 ml (b) 714 ml oil, 286 ml vinegar   4(a) 3.3 cm
(2SF) (b) 675 Newtons   5 £2.13 to the nearest penny, 175 cm   6 £15 000,
£6000, £3000   7 124°, 178°, 59° to nearest degree   8 (a) 765 kg (b) 35 000 1
(2SF)   9(a) 8 men (b) 13 h 20 min

Percentages

1(a) 0.05, 5% (b) 0.7, 70% (c) 0.8, 80% (d) 0.375, 37½% (e) 0.67 (2SF), 67%
(2SF) (f) 0.417 (3SF), 41.7% (3SF) (g) 1.5, 150%   2(a) 0.2, $\frac{1}{5}$ (b) 0.75, $\frac{3}{4}$
(c) 0.065, $\frac{13}{200}$ (d) 0.88, $\frac{22}{25}$ (e) 1.06, $\frac{53}{50}$ (f) 0.125, $\frac{1}{8}$ 3(a) 54 km (b) 67p to
nearest penny (c) 20 002 votes (d) £189.50 (e) 4200 ants (f) 0.017 mm
4(a) 215 people (b) 64.75 kg (c) £401.35 (d) 90 million per year   5(a) 18.75%
(b) 8.8% (c) 8.3% (2SF)   6(a) 73.9 kg (b) 237 km² (c) £13 703.70
7 1.06 m, 33%   8 542, 2000

Buying and Selling

1 £7.34   2 £10.90   3 £49.45   4 £260   5(a) £41.25 (b) £71.75 (c) £50.50
(d) £73.75   6 Dick £30, Harry £25.50   7 £187.39   8(a) £1.25 (b) £3.75
9 69p   10 £15.86   11 Fresh beans 48p   12 22% (2SF), £26.22

Saving and Borrowing

1(a) £27 (b) £4.57 (c) £5505.50 (d) £285.75   2(a) £285 (b) £298.54
3 £8010.40   4 £320, £35   5 £4275, £55 959   6 9.5% p. a.   7 £5558.40,
16.7% (1DP)

Earning a Wage

1(a) £92.40 (b) £110.55 (c) £128.70   2(a) £59.25 (b) £92.82 (c) £80.03
(d) £93.28   3(a) £386.43 (b) £516.41 (c) 781.55   4 £8124.48   5 £492.33
6 £3645.80, £70.11   7 £124.32, £37.96

Reading Tables

1(a) 59.4 cm × 84.1 cm (b) 28.2 in (c) A4   2(a) £427 million (b) £1379 million
3(a) 0.69 (2DP) (b) 1 h 41 min   4(a) 91.44 cm (b) 454.6 l (c) 2.20 lb (2DP)
(d) 31.08 mph (2DP)   5(a) 100 m² (b) 200 gal (c) 5 kg (d) 7 ft (e) 50 mph
(f) 1 l.

Time
1 8 h 9 min **2** 1911 **3** 4 h 40 min, 7.20 am **4** 1 h 15 min **5** 4 min 18 s, 1 min 4.5 s **6**(a) 1933 (b) 12 min (c) 0833, 0945, 1400, 1612 **7**(a) 26th March '86 (b) 3rd April '86 (c) 15th March '86, 9th March, 13th April, 11th May '86 **8**(a) 12.53 pm (b) 12.26 am

Units
1(a) 0.023 m (b) 56000 g (c) 750 m (d) 0.05 g (e) 225 cm^3 (f) 0.034 m^3 (g) 17 ft (h) 139 lb (i) 0.25 pt **2** 0.255 l **3**(a) 2635.54 kg (b) 61 450 kg (c) 28.8775 kg **4**(a) 217 000 m^2 (b) 0.217 km^2 (c) 21.7 ha **5**(a) 2.5 kg (b) 7.5 cm (c) 1.5 km (d) 900 l **6**(a) 52.77 pt (b) 13.64 l (c) 189.5 ml **7**(a) 1070 lb (b) 76 st 8 lb **8**(a) 7761 pesetas (b) 10 725 guilders (c) 109 657 lira **9** £14 340 million **10** 43 p **11**(a) 98.4°F (b) 22.5°C **12** 147

Length
1 30 ft **2** 4 m, 5 m **3**(a) 7 m × 7 m (b) 4 m × 3 m (c) 110 m^2 (d) 128 m^2 (e) 16% **4**(a) 800 km (b) 1000 km (c) 2350 km **5**(a) 4 mm (b) 1.5 mm (c) 2 mm **6**(a) 75 m (b) 275 m (c) 5 m (d) 90 m (e) 100 m (f) 210 m **7**(a) 4.5 cm (b) 2.25 cm (c) 1.13 mm (d) 9 cm (e) 22.5 cm (f) 27 cm

Scale
1 17.5 m, 4.38 m **2** 12 cm, 19 km, 0.8 cm, 0.25 km, 0.48 cm **3**(a) 750 m (b) 3.5 km (c) 100 m (d) 20 km **4** 210 mm, 9 **5** 10 000, 1 × 10^{18} **6** 16 p **7** 844 ml **8** 125 000 **9**(a) 3 m × 2.25 m, 3.75 m × 3.75 m, 3 m × 1.5 m (b) 25.3 cm^2

Angle
1(a) 240° (b) 36° (c) 15° (d) 135° (e) 255° (f) 600° (g) 285° **2**(a) $q = 137°$ (b) $r = 148°$ $s = 32°$ $t = 51°$ (c) $u = 77°$ $v = 13°$ $w = 77°$ (d) $x = 58°$ $y = 58°$ $z = 122°$ (e) $a = 37°$ $b = 49°$ $c = 94°$ **3**(a) 18° (b) 50° (c) 35° (d) 125° **4** 15 km 106°, 20 km 245°, 24 km 273°

Perimeter
1 154 m **2**(a) 38.5 cm (b) 14.6 km (c) 1770 cm (3SF) **3** 5.0 m **4** 28 m, £301.50 **5** 70 cm

Circles
1(a) 1040 mm, 85 500 mm^2 (b) 54.6 m, 238 m^2 (c) 78.5 km, 491 km^2 (d) 0.0785 cm, 0.000 491 cm^2 (e) 9.42 in, 7.07 in^2 **2** 66 m **3** 7.8 m **4** 400 m, 9260 m^2, 18.8 m **5** 9140 m^2, 384 m **6** 11 500 mm

Area
1(a) 1200 m^2 (b) 1000 m^2 (c) 1200 m^2 (d) 700 m^2 (e) 7000 m^2, 10% **2**(a) 689 ft^2 (b) 4.7 m^2 (c) 19 mm^2 (d) 1200 cm^2 **3** 38.6 mm^2 **4**(a) 72 cm^2 (b) 412 cm^2 (c) 484 cm^2 (d) 300 **5** Wym. 91 000 sq mile, 3.6; Col. 104 000 sq mile, 21; Kan. 80 000 sq mile, 28; Neb. 83 000 sq mile, 18

Nets and Surface Area
1 9.5 m^2, £46 **2**(a) 4.5 m^2 (b) 2.5 m^2 (c) £1.12, 0.22 **3** 210 m^2, £525, £65 **4**(a) 17 m^2 (b) 47 m^2 (c) 17 m^2, 2 trays, 3 tins, £160 **5** cuboid, £3.10, £3.07, £2.80

Answers 174

Volume
1 131 000 m^3 2(a) 145 cm^3 (b) 138 000 mm^3 3(a) 385 cm^3 (b) 23 100 cm^3
(c) 60   4 3.8 × 10^{-5} m^2, 1.2 × 10^{-4} m^3, 3 h 22 min   5(a) 55 cm^2 (b) 340 cm^3,
28 sweets, 97 cm^3 6(a) 8.18 m^3 (b) 8180 l, 2.01 m   7(a) 1 g   (b) 1 kg
(c) 2 mg   8 25.0 l, 31.7 l, 38.5 l, 193 l   9 18 000 cubic tukits, 17 000 cubic tukits

Shape and Symmetry
1(a) 6, 6   (b) 7, 7   (c) 8, 8   2(a) $x = 79°$, $y = 101°$ (b) $x = 48°$, $y = 61°$,
$z = 289°$ (c) $x = 31°$, $y = 62°$, $z = 62°$, isosceles triangle, trapezium, rhombus
3(a) 120°   (b) 60°   (c) 135°   (d) 150°

Solids
1(a) 7, 15, 10   (b) 7, 12, 7   (c) 8, 18, 12   3(a) 3   (b) 5   (c) 4   (d) infinite

Constructions and Loci
4(a) circle, centre A, radius 3 cm   (b) half plane containing B bounded by the
mediator of AB (c) two half lines from B at 30° to BA each side   (d) circle,
diameter AB (e) arc of circle, centre A, radius 5 cm, lying inside circle, centre B,
radius 3 cm   8 20 cm

Prove it
1(a) $a = 90°$, $b = 58°$, $c = 58°$, $d = 8°$ (b) $p = 42°$, $q = 48°$, $r = 48°$, $s = 54°$
3 8 cm

Trigonometric Ratios
1 $u = 46.5$ cm, $v = 47.9°$, $w = 5.88$ cm, $x = 131$ km, $y = 136$ km, $z = 62.4°$
2 17.2 mm, 41.8°, 48.2°   3 34.3 cm, 55.3°   4 46.9 km, 746, 71.3°

Two-dimensional Problems
1 0.7° (1SF)   2 695 m   3 136 m   4 218 m   5 15.9 km, 15.9 km, 161 kph
6 64.8 m, 48.8 m   7 4.91 m, 21.2°   8 6.65 m, 33.5 m

Three-dimensional Problems
1 7 ft 10 in, 120 sq ft   2(a) 35.2 m   (b) 49.7 m   (c) 32.4°   3(a) 68.2°
(b) 19.8 cm   (c) 60.5°   4(a) 38°   (b) 2.6 m   (c) 5.5 m   5(a) 16.8 m   (b) 40 m^2

Making Waves

1(a)
x	0	30	60	90	180	210	240	270	360
y	1	2.5	3.6	4	1	−0.5	−1.6	−2	1

(b)
x	0	22½	45	67½	90
y	4	6	8	6	4

2
t	0	2	4	6	8	10	12	18	24
d	13	12.3	10.5	8	5.5	3.6	3	8	13

(a) 5.5 ft   (b) 12 noon   (c) before 0736 and after 1624

3
t	0	0.2	0.4	0.6	0.8	1.0	1.2	1.4	1.6	1.8	2.0	2.2	2.4	2.6	2.8	3.0
l	34	46	52	50	40	28	18	16	22	34	46	52	50	40	28	18

(a) 52 cm   (b) 16 cm   (c) 1.8 s

Algebra
1(a) −1 (b) 1 (c) 5 (d) −5 (e) −10 (f) 6 (g) −4 (h) 13 (i) 0
(j) −6 (k) 9 (l) −2 (m) −18 (n) $-\frac{2}{3}$ (o) −2 (p) $-\frac{3}{5}$ **2**(a) −18
(b) −11 (c) −1 (d) 11 (e) −1 (f) 371 **3**(a) −49 (b) 819 (c) 448
4(a) $18x + 11y$ (b) $17b − 9a$ (c) $11a^2 − 5ab$ (d) $3m^2 + 2mn$ **5**(a) $(13x − 2)/6$
(b) $(22x + 1)/15$

Factorisation
1(a) $p(2q − 1)$ (b) $3(4q^2 + 2pq − p)$ (c) $2(r + 3s)$ (d) $3xy(x + 3y − 2)$
(e) $5(a − 2b^2 + 4a^2)$ (f) $14bc(4a − 3b)$ **2**(a) $(x − 7)$ (b) $(x − 2)(x + 3)$
(c) $(2x + 5)(x − 8)$ (d) $(2x − 3)(2x + 3)$ **3**(a) $(x + 3)(x + 3)$ (b) $(x − 4)(x − 3)$
(c) $(x − 3)x + 4)$ (d) $(x + 3)(x − 5)$ (e) $(2x + 1)(x + 5)$ (f) $(3x − 1)(x − 4)$
(g) $(2x − 3)(x + 1)$ (h) $(4x − 3)(x + 2)$ (i) $(3x + 2)(2x − 9)$

What is a Function?
1(a) 7 (b) 1 (c) −5 (d) 15 (e) 5 **2** $2.7x, 2.7x + 1.8$ (a) 64.16
(b) $(2.7x + 1.8)^2$, $-\frac{2}{3}$ **3**(a) $-\frac{1}{3}$ (b) $15\frac{1}{2}$ **4** $x = -2, x = 6.3$

Conversion Graphs
1(a) £28.40 (b) 2800 ft³ (c) £9.20 **2**(a) £15 (b) £26.50 (c) 38 (d) £20
(e) 350 **3**(a) £25 (b) £25 (c) £40 (d) £85 (e) £32.50 (f) 4 h 20 min

Travel Graphs
1(a) 30 km/h (b) 102 km/h (c) 100 km/h (d) 6 km/h (e) 0.5 cm per min
(f) 60 mile/h **2**(a) 6 h (b) $\frac{1}{2}$ h (c) 5 h 43 min (d) 12 min (e) 3 min
(f) 11 min **3**(a) 450 km (b) 210 km (c) 3000 km (d) 620 mile (e) 249 km
(3SF) **4** 180 km, 60 km/h **5**(a) 100 km (b) $\frac{1}{2}$ h (c) 50 km/h (d) $33\frac{1}{3}$ km/h
7 130 m **8**(a) (i) 150 km (ii) 200 km (b) $\frac{1}{2}$ h (c) (i) 1330 (ii) 1700 (d) 2 h
(e) (i) 80 km/h (ii) 87 km/h **9** $\frac{1}{2}$ km, 1 min **10**(a) 27 s after Mary starts
(b) 5 lengths (c) 4

Plotting Functions
1(a) 0, 75, 150, 225 (b) 225 litres (c) (i) 64 litres (ii) 32 cm **2**(a) $x = 1.7$
(b) $x = 6.6$ or -0.6 (c) $x = 1.35$ **3**(a) 17 s (b) 48 **4**(a) £14 (b) £26
5 $C = 10t + 15$, £10

Interpreting Graphs
1(a) the first graph (b) the third graph (c) the second graph **2** Fill, pause, get
in, soak, pull plug, refill, soak, empty **3**(a) 7 ms⁻¹ (b) $-1\frac{1}{4}$ ms⁻¹ (c) 68 m
4 5 ms⁻¹, $\frac{1}{2}$ ms⁻², 20 ms⁻¹, 1 ms⁻², 30 ms⁻¹, 0, 800 m, 75 s, 225 m

Composite Functions
1(a) 7 (b) 0 (c) 3 (d) −48, $fg(x) = 7 − 4x^2$, $gf(x) = -16x^2 − 24x − 8$ **2**(a) −9
(b) +7 (c) ×4 (d) ÷2 (e) square (f) +/− **3**(a) $7(x + 2)$ (b) $x^2/3$
(c) $9(3 − x)$ **4**(a) $(x − 2)/5$ (b) $2/(3 − x), x \neq 3$ (c) not possible **5**(a) h (b) 3, 1
(c) $x^2 + 2x − 3$, $x^2 − 4x + 3$ **6**

x	−1	0	1	2	3	4
3x+1	−2	1	4	7	10	13

$f^{-1}(x) = (x − 1)/3$, reflection in $y = x$

Linear Functions

2(a) $m = 1, c = -3$, (b) $m = \frac{1}{2}$, $c = 2\frac{1}{2}$, (c) $m = -\frac{2}{3}$, $c = 4$, (d) $m = -3$, $c = 5$, (e) $m = -4$, $c = 7$, 3(a) 4 (b) -3 (c) $-\frac{2}{5}$ (d) $\frac{7}{5}$
4(a) $y = 2x - 7$ (b) $y = 5 - x$ (c) $y = -(x+1)$ (d) $2y + x = 0$ 5 (12, 0), (0, 5), $-\frac{5}{12}$, 13 6 AB $4y = x$, BC $y = 17 - 4x$, CD $4y = 17 + x$, AD $y + 4x = 0$

Linear Equations

1(a) $x = 8$ (b) $x = -5$ (c) $x = 4$ (d) $x = -4$ (e) $x = 11$ (f) $x = 7$
(g) $x = -2$ 2(a) $x = 4$ (b) $x = 3$ (c) $x = 5$ (d) $x = 5$ (e) $x = 6$
(f) $x = 10$ (g) $x = 31$ 3(a) $x < 5$ (b) $x < 9$ (c) $x > -3$ (d) $x > 8$
(e) $x < 2$ (f) $x < -\frac{2}{3}$ (g) $x < -\frac{15}{2}$ (h) $x < 2$ 4 3 5(a) $-\frac{25}{4}$ (b) $x > \frac{1}{3}$
6 52

Problem Solving with Algebra

1 $x - 7$, 33 p 2 $27 - b$, 9 3 $x = (y + 7)/3$ 4(a) -16 (b) 4 (c) 7 ms^{-2}
5 $n + 1$, 13 6 $m + 17$, $m + 47$, 29, 46, 76 7(a) $3 - 2x$, $4 - 2x$
(b) $(3 - 2x)(4 - 2x)$ (c) $14x - 4x^2$ 8 40, 37, 4 9 $12 \text{ m} \times 18 \text{ m}$ 10(a) $18x$ p
(b) $24x$ p (c) $42x$ p, 33

Simultaneous Equations

1(a) $x = 3.6, y = 1.7$ (b) $x = 2.9, y = 1.3$ (c) $x = -0.7, y = 2.1$ (d) $x = 2.2$, $y = 0.7$ 2(a) $x = 3, y = 1$ (b) $x = 4, y = 2$ (c) $x = 5, y = -2$ (d) $x = 1$, $y = 4$ (e) $x = -1, y = -2$ (f) $x = 7, y = -3$ (g) $x = 2, y = 5$ (h) $x = 1$, $y = 1$ 3 $5x + 3y = 17$, $3x + 2y = 10.5$, £2.50, £1.50 4 $x = 1, y = 4$ or $x = 2, y = 7$ 5 $x = -1$ or 5 6(a) $x = 0, y = 0$ or $x = 2, y = 4$ (b) $x = 1$, $y = -3$, (c) $x = 2, y = 7$ or $x = -2, y = -1$ (d) $x = 3.3, y = 2.3$ or $x = -4.3$, $y = -5.3$ 7 10, 6, 4, 18, $x = 4$ or -2

Quadratic Equations

1(a) $x = 3$ or -1 (b) $x = -2\frac{1}{2}$ or 4 (c) $x = \frac{1}{3}$ or $-\frac{1}{5}$ (d) $x = 8$ or 1 (e) $x = 3$ or -5 (f) $x = 5$ or -1 (g) $x = 9$ or -7 (h) $x = 4$ or -11 (i) $x = 7$ or -7
(j) $x = 1$ or $\frac{1}{2}$ (k) $x = -3$ or $\frac{5}{3}$ (l) $x = -\frac{4}{3}$ or $\frac{5}{2}$ 2(a) 6.8 or 1.2 (b) -5.7 or 2.7 (c) 4.2 or -1.2

3

x	-3	-2	-1	0	1	2	3	4
$f(x)$	23	11	3	-1	-1	3	11	23

$-0.4, 1.4, -0.37, 1.37$

4(a) 1.2 -4.2 1.19 -4.19 (b) 0.6 3.4 0.59 3.41 (c) 2.4 -0.9 2.35 -0.85
(d) 4.2 -0.2 4.24 -0.24 (e) 1.2 -0.5 1.22 -0.55 (f) 6.4 0.6 6.37 0.63 5 $x = -1$ or 5 6 $x = 1$ or -7 (b) $x = -\frac{1}{2}$ or $\frac{11}{2}$ (c) $x = -2$ or 7 (d) $x = 2.4$ or 0.3 (e) $x = 2.4$ or -4.6 (f) $x = 4.1$ or 1.2 7(a) $x = 4$ or -4 (b) $x = 3$ or -2 8(a) $x = 5.3$ or 1.7 (b) $x = 1.9$ or -5.9 (c) no soln. (d) no soln.
(e) $x = 4$ or $-\frac{1}{2}$

Linear Inequalities

1(a) $-1 \leqslant x \leqslant 3$ (b) $-2 < x < -1$ or $1 \leqslant x \leqslant 3$ (c) $x < 0$ or $1 \leqslant x$
(d) $-1 \leqslant x < 0$ or $1 < x \leqslant 2$ or $3 < x$ 3(a) $-1, 0, 1$ (b) 1, 2, 3 (c) $-4, -3$, -2 (d) 1 4(a) (2, 1) (3, 1) (4, 1) (2, 2) (3, 2) (4, 2) (b) (1, 1) (2, 1) (1, 2) (c) (2, 2) (2, 3) (2, 4) (3, 3) (3, 4) (4, 4) (d) (3, 5) (4, 4) (4, 5) (e) (1, 0) (0, 1) (1, 1) (1, 2)
5 $x + y \leqslant 7$ $15x + 10y \leqslant 90$, $x \geqslant 2, y \geqslant 2$, (4, 3), £43 6(a) £1 (b) £1.70 7 40 white, 80 brown

Collecting and Sorting Data
1(a) Street (b) $\frac{11}{36}$ 2(a) Tally (b) 12, 8, 26, 9, 15, 4, 7 (c) 81
3 6.7 cm (2SF)   4 12, 6.5   5 125 cm   6 4, 5, 9   7(a) (i) 49.8 (3SF) (ii) 50
(b) 0.64

Charts
1 48, 22   2(a) Angles × 20 (c) 2 (d) 1.7 (2SF)   3(a) 225 (b) 1550
(c) 260 (2SF) (d) £14 800 (2SF) (e) Xmas is near   4 £32   5 Angles × 10, 22, 75 p

Pictorial Information
1(a) 11 750 (b) 95% (c) $14\frac{3}{4}$ 2(a) No scale on production axis (b) Turnover may not increase at the same rate   3 6400, 14 400   4 Sales scale drawn to make tiny sales increase look dramatic   5(a) 11 kg (b) 82 cm   6(a) 31.7 cm (b) 33.8 cm (c) 9.4 kg, outside range of data   7(a) 31 in (b) 79 cm, long

Grouped Data
1 19.7 (3SF)   2(a) 5 (b) 0 (c) 1 (d) 2 (e) 1.5 (2SF)   3(a) £6500 (b) £7250 (c) £10 800 (3SF), although mean is £10 000+, the modal salary is £6500 and half the workforce earn £7250 or less

Taking a Chance
1 $\frac{2}{3}$ 2(a) $\frac{2}{9}$ (b) $\frac{5}{9}$ 3 0.85   4(a) $\frac{1}{2}$ (b) $\frac{1}{3}$ 5 $\frac{11}{28}$ 6 $\frac{1}{2}$ 7(a) $\frac{1}{4}$ (b) $\frac{1}{52}$ (c) $\frac{3}{13}$ 8(a) $\frac{1}{4}$ (b) $\frac{3}{8}$ 9 THH TTH THT HTT, $\frac{3}{8}$ 10 $\frac{9}{32}$ 11(a) $\frac{7}{10}$ (b) $\frac{3}{20}$ (c) $\frac{2}{5}$ 12 $\frac{4}{5}$, $\frac{80}{99}$

Combined Events
1 $\frac{1}{12}$, $\frac{1}{3}$ 2(a) $\frac{2}{35}$ (b) $\frac{27}{70}$ 3(a) $\frac{1}{9}$ (b) $\frac{19}{36}$ 4(a) $\frac{3}{7}$ (b) $\frac{15}{91}$ (c) $\frac{4}{13}$ 5(a) 40 (b) $\frac{1}{16}$ 6(a) $\frac{1}{5}$ (b) $\frac{4}{125}$ 7 $\frac{1}{6}$, $\frac{7}{12}$ 8 $\frac{11}{21}$ 9(a) $\frac{3}{8}$ (b) $\frac{3}{8}$ 10(a) $\frac{1}{2}$ (b) (i) $\frac{1}{6}$ (ii) $\frac{1}{12}$

Enlarge it
2(c) translate $\begin{pmatrix} 3 \\ 0 \end{pmatrix}$ 3(a) translate $\overrightarrow{AY}$ (b) rotate 180° about midpoint of XY (c) Enlarge, centre A, SF 2   4 3, (3, 5)

Translation, Reflection, Rotation
4 reflect in AC, rotate 180° about B, translate $\begin{pmatrix} 3 \\ 0 \end{pmatrix}$ 5(a) (5, 0) (7, 2) (8, 1) (b) (5, 0) (3, −2) (2, −1)

Combined Transformations
1 reflect in $y = 0$ 2(b) (i) Translate $\begin{pmatrix} 6 \\ 2 \end{pmatrix}$ (ii) Reflect in x axis   3(e) T (f) S (g) I (h) I   4(a) $y = 1$ (b) $x = 0$ (c) $y = -1$ (d) $y = 2$ (e) $x = 1$ (f) $x = -1$

Vectors

1(a) $-w$ (b) $z-w$ (c) $x+y+z-w$ (d) $\frac{1}{2}(w+x+y+z)$ 2(a) e (b) a (c) $-a$ (d) c 5 $\frac{1}{2}a-b$, $(a-2b)/3$, $2:1$ 6(a) $\begin{pmatrix}4\\2\end{pmatrix}$ $\begin{pmatrix}1\\4\end{pmatrix}$ $\begin{pmatrix}3\\3\end{pmatrix}$ $\begin{pmatrix}2\\-1\end{pmatrix}$ $\begin{pmatrix}-4\\1\end{pmatrix}$

(b) $\begin{pmatrix}1\\2\end{pmatrix}$ $\begin{pmatrix}-1\\-1\end{pmatrix}$ $\begin{pmatrix}2\\1\end{pmatrix}$ (c) $\overrightarrow{AB}$ c, $\overrightarrow{EF}$ b, (d) a c $\overrightarrow{GH}$ $\overrightarrow{CD}$ $\overrightarrow{IJ}$ 8(a) $\begin{pmatrix}3\\2\end{pmatrix}$

(b) $\begin{pmatrix}6\\7\end{pmatrix}$ (c) $\begin{pmatrix}0\\11\end{pmatrix}$ (d) $\begin{pmatrix}-4\\2\end{pmatrix}$ (e) $\begin{pmatrix}-4\\4\end{pmatrix}$ (f) $\begin{pmatrix}8\\-2\end{pmatrix}$ 9(a) $\begin{pmatrix}7\\1\end{pmatrix}$ (b) $\begin{pmatrix}-10\\50\end{pmatrix}$

(c) $\begin{pmatrix}-4\\-2\end{pmatrix}$ (d) $\begin{pmatrix}22\\62\end{pmatrix}$ (e) $\begin{pmatrix}-6\\27\end{pmatrix}$ (f) $\begin{pmatrix}28\\10\end{pmatrix}$ (g) 6.4 10(a) $\begin{pmatrix}3\\-4\end{pmatrix}$, 5

(b) $(-1, -6)$ 11(a) $\begin{pmatrix}6\\-2\end{pmatrix}$, $\begin{pmatrix}2\\4\end{pmatrix}$ (c) $\begin{pmatrix}6\\-2\end{pmatrix}$

Matrices

1 3, -4, -3, -5 2 $\begin{pmatrix}1 & -1 & 2\\-6 & 9 & -12\\5 & -6 & 10\end{pmatrix}$, $\begin{pmatrix}7 & 5\\14 & 13\end{pmatrix}$

3(a) $\begin{pmatrix}-1 & 1\\0 & -4\end{pmatrix}$ (b) $\begin{pmatrix}-1 & 2\\0 & -4\end{pmatrix}$ (c) $\begin{pmatrix}-3 & 3\\2 & 0\end{pmatrix}$ (d) $\begin{pmatrix}-2 & 8\\1 & -1\end{pmatrix}$

(e) & (f) $\begin{pmatrix}-3 & 3\\4 & 0\end{pmatrix}$ 4 $p=1, q=-5$ 5(a) 2×3, 2×3 (b) $3\times n$, $2\times n$,

$n \in \mathbb{N}$ 6 $S=\begin{pmatrix}4 & 3 & 8 & 0\\2 & 6 & 0 & 12\\0 & 5 & 0 & 20\end{pmatrix}$, $P=\begin{pmatrix}23\\31\\3\\9\end{pmatrix}$, $SP=\begin{pmatrix}209\\340\\335\end{pmatrix}$, £2.09, Kevin, 884,

Total amount spent £8.84 7 $\begin{pmatrix}60 & 25 & 20 & 0\\70 & 0 & 0 & 30\\0 & 30 & 10 & 25\end{pmatrix}$, $\begin{pmatrix}12\\10\\9\\9\end{pmatrix}$, 1150 kg, 1110 kg,

615 kg 8(a) $\begin{pmatrix}0 & 0\\0 & 0\end{pmatrix}$ (b) $\begin{pmatrix}3 & 8\\-8 & -5\end{pmatrix}$ (c) $\begin{pmatrix}-3 & -8\\8 & 5\end{pmatrix}$

Route Matrices

1 $\begin{pmatrix}0 & 2 & 2\\2 & 0 & 1\\1 & 1 & 1\end{pmatrix}$, $\begin{pmatrix}6 & 2 & 4\\1 & 5 & 5\\3 & 3 & 4\end{pmatrix}$, 4 2 (b) Yes (c) $B \to M \to L \to S$

3(a) $\begin{pmatrix}1 & 1\\1 & 1\\0 & 1\end{pmatrix}$ (b) $\begin{pmatrix}1 & 0 & 1\\0 & 1 & 1\end{pmatrix}$ (c) $\begin{pmatrix}1 & 1 & 2\\1 & 1 & 2\\0 & 1 & 1\end{pmatrix}$, 2 4 1. Enid, 2. Fanny, all rest equal 3rd

Matrix Transformations

2(a) 90° rotate about (0, 0) (b) $-90°$ rotate about (0, 0) (c) Enlarge, centre (0, 0), SF2 (d) Reflect in $y+x=0$ (e) Enlarge, centre (0, 0), SF -3

3(a) $\begin{pmatrix}0 & -1\\-1 & 0\end{pmatrix}$ (b) $\begin{pmatrix}-1 & 0\\0 & -1\end{pmatrix}$ (c) $\begin{pmatrix}0 & 5\\-5 & 0\end{pmatrix}$ 4 $\begin{pmatrix}3 & -4\\4 & 3\end{pmatrix}$, $(-7, 24)$, $(0, -1)$

Enlarge centre (0, 0) SF5. Rotate 53° about (0, 0), $\frac{1}{25}\begin{pmatrix}3 & 4\\-4 & 4\end{pmatrix}$